The Philosophy of Cognitive Science

To Julie, Maya and Theo

The Philosophy of Cognitive Science

M. J. CAIN

polity

First published in 2016 by Polity Press

Polity Press
65 Bridge Street
Cambridge CB2 1UR, UK

Polity Press
350 Main Street
Malden, MA 02148, USA

ISBN-13: 978-0-7456-4656-5
ISBN-13: 978-0-7456-4657-2 (pb)

A catalogue record for this book is available from the British Library.

Library of Congress Cataloging-in-Publication Data

Cain, M. J.
  The philosophy of cognitive science / Mark Cain.
      pages cm
  Includes bibliographical references and index.
   ISBN 978-0-7456-4656-5 (hardcover : alk. paper) -- ISBN 0-7456-4656-5 (hardcover : alk. paper) -- ISBN 978-0-7456-4657-2 (pbk. : alk. paper) -- ISBN 0-7456-4657-3 (pbk. : alk. paper) 1. Philosophy and cognitive science. 2. Philosophy of mind. I. Title.
  BD418.3.C34 2015
  128'.2--dc23
                              2015008643

Typeset in 9.5 on 12 pt Utopia by
Servis Filmsetting Ltd, Stockport, Cheshire
Printed and bound in Great Britain by CPI Group (UK) Ltd, Croydon, CR0 4YY

For further information on Polity, visit our website: politybooks.com

Contents

Acknowledgements   vi

1 Cognitive Science and the Philosophy of Cognitive Science   1

2 Representation and Computation   25

3 Modularity   66

4 Concepts   93

5 Language   123

6 The Brain and Cognition   161

Conclusion   195

Notes   199
References   207
Index   219

Acknowledgements

Writing this book has taken much longer than I anticipated when I first sketched its outline several more years ago than I care to dwell upon. I therefore need to thank my editors Emma Hutchinson and Pascal Porcheron at Polity for their continued patience, support and encouragement. Thanks also to my philosophy colleagues at Oxford Brookes University, Steve Boulter, Constantine Sandis, Dan O'Brien and Bev Clack. Rumour has it that they find many of my philosophical views slightly bonkers, but that hasn't stopped them providing a warm and stimulating environment to work in. I've also had the pleasure of supervising the PhD research of Dana Vilistere, Annie Webster, Marcus Westberg, Andrew McIntyre, Colin Beckley, Linda Fisher-Hoyrem, Esther Shallan and Istvan Zardai, all of whom have provided very helpful feedback on my developing ideas. Massive thanks to my daughter Maya Reid-Cain for helping me wrestle with the various diagrams used in this book and for keeping her disdain at my IT ineptitude in check. Most of all thanks to my wife Julie and children Maya and Theo for all their love and support. This book is dedicated to them.

1 Cognitive Science and the Philosophy of Cognitive Science

1 Introduction

This is a book about the philosophy of cognitive science. That topic immediately raises two questions: what is cognitive science and what is the philosophy of cognitive science? In one respect the answers to these questions are obvious: cognitive science is the scientific study of cognition, and the philosophy of cognitive science is that branch of philosophy that addresses philosophical questions generated by the scientific study of cognition. But these answers are hardly illuminating as they raise a number of subsidiary questions. What is cognition? What are the key assumptions and methods adopted by those who attempt to study cognition scientifically? How does cognitive science relate both to other sciences and to our commonsense understanding of ourselves as minded agents? When and how did cognitive science emerge as a distinct discipline? How does the philosophy of cognitive science relate to other branches of philosophy such as the philosophy of mind and the philosophy of science? In this chapter I will address these general questions and so provide the foundations for the more specific discussions of the later chapters.

2 Cognition

What exactly is cognition? In order to answer this question it is helpful to begin from a commonsense perspective. According to commonsense – or at least the commonsense perspective of most twenty-first-century Westerners – human beings can be distinguished from inanimate physical objects in having a mind. What, according to commonsense, is involved in our having minds? Here is a by no means exhaustive list: we think; we perceive the external world by means of our senses; we experience pain and other sensations when our bodies are appropriately stimulated; we experience moods such as depression and light-heartedness; we experience emotions such as anger, joy and jealousy; we are conscious both that we think and feel and how we specifically think and feel; we act on the basis of our decisions and intentions, which in turn often reflect how we perceive the world to be, what we think and what we want; when we

1

act we are conscious of what we do without having to rely on our external sense organs; we recollect our past thoughts, actions and experiences; we imagine particular scenarios; and so on.

In so describing the activities that are central to human mentality from the commonsense perspective I did not use the term 'cognition'. But if one were to ask which of the above activities were best described as involving cognition, most people would answer 'thinking'. So one might say that, to a first approximation, cognition is thinking. But what exactly is thinking? Thinking is a mental process or activity that results in having a thought. Such thought processes range from the intellectually demanding and abstruse (as when one thinks when doing philosophy) to the banal and everyday (as when I address the question of whether I have enough time before my next lecture to buy a coffee). Some concern matters of great importance, others not. Some issue in thoughts that have an immediate impact on action and others don't. Some issue in thoughts that become deeply entrenched and influence many subsequent thought processes, while others issue in thoughts that are fleeting and inconsequential. Some concern counterfactual or hypothetical matters (would I have been late for my lecture had I queued for a coffee?) whereas others are straightforwardly factual.

If thinking is a mental process that issues in a thought, then what exactly is a thought? In everyday talk the term 'thought' is most commonly used to refer to the mental state of considering a particular hypothesis or answer to a question or that of committing oneself to a particular hypothesis or answer. Such mental states are, or are closely related to, beliefs. For when one considers a particular answer to a given question one is paving the way to holding a particular belief, and when one commits oneself to a particular answer one has thereby acquired a belief. Beliefs are examples of what philosophers call propositional attitudes, as believing that p involves adopting the belief attitude to the proposition p. But there are propositional attitudes that are not beliefs, for, just as one can believe that p, one can desire that p, intend (to make it the case) that p, hope that p, fear that p, expect that p, and so on. Consequently, although it slightly strains everyday usage, one might say that thoughts are propositional attitudes, so that thinking is the mental process that results in the acquisition of a propositional attitude, be it a belief, a desire, an intention, or whatever.

In virtue of being relations to propositions, propositional attitudes have meaning or semantic properties. They are therefore akin to declarative sentences of a natural language. Just as such sentences are about particular things (or types of thing) or states of affairs and represent them as being a particular way, so do propositional attitudes. For example, my belief that aardvarks eat termites is about aardvarks and represents them as being termite eaters. What a propositional attitude is about and how it represents its object are elements of its meaning, but philosophers tend to use the term 'content' when talking about the meaning of propositional attitudes. So, for example, the sentence 'aardvarks eat termites' has a particular meaning, and the belief that aardvarks eat termites has a particular (corresponding) content. There is a further commonality between declarative sentences and propositional attitudes. Sentences are made up of simpler components, namely words, and the meaning of a sentence is a product of the meaning of the words that belong to it and the way they are put together (the syntactic structure of the sentence).[1] It is natural to think that something similar is true of propositional attitudes. The simpler components of propositional attitudes are concepts, and the content of a propositional attitude is a product of its constituent concepts and the way they are put together. Hence, one cannot form the belief that aardvarks eat termites without employing each of its constituent concepts.

I have suggested that, to a first approximation, cognition is thinking. Allied with the idea that thinking is a mental process that results in the formation of propositional attitudes, the implication might seem to be that much of what goes on in the mind falls outside of the domain of cognition. This is because philosophers standardly distinguish between propositional attitudes and other mental states such as sensations, emotions and perceptual experiences. However, even if we want to uphold such a distinction (say, on the grounds that such mental states differ from propositional attitudes in that they have an essential phenomenal, qualitative or 'what it is like' aspect), it would perhaps be a little rash, for several reasons, to conclude that only propositional attitudes and the processes that generate them count as cognitive. First, consider perception. Perception is a process where the external world stimulates an individual's sensory organs, resulting in a perceptual experience. In the case of vision, light reflected off external objects is focused onto the retina, a light-sensitive surface at the back of the eye that sets off a mental process which results in a visual experience. Even if, as many philosophers have thought, perceptual experiences are very different from propositional attitudes in not involving the deployment of concepts and in having an intrinsic qualitative character, they are like them in a key respect. For perceptual experiences are representational in the respect that they are typically about objects located in the external world and represent them as being a particular way. For example, my current visual experience was not only caused by the computer directly in front of me but is also about that computer and represents the outer world immediately before me as containing a grey, rectangular object. In virtue of their being representational, it seems reasonable to think of perceptual states as belonging to the domain of cognition if propositional attitudes are paradigmatic cognitive states.

A second point relates to the nature of the mental process of thinking.

Typically, when one acquires a propositional attitude by means of thinking, a process takes place in one's mind that is extended in time. This process has stages each of which involves moving from one or more propositional attitudes to another, where the earlier propositional attitude(s) in each stage justify the later. Here is an example. Feeling a little de-energized, I wonder if I have enough time to buy a coffee before my next lecture. I start from the beliefs that it will take me ten minutes to walk from the café to the lecture hall and that, given the current length of the queue, it will take me five minutes to purchase a coffee. Given that my lecture starts at 1.00 p.m., I infer that I must leave the café by 12.45 p.m. at the latest to be on time. I look at my watch and come to believe that it is 12.50 p.m. and draw the implication that I cannot buy a coffee and arrive at my lecture on time. Given my strong desire not to be late, I decide to give coffee a miss.

When we think, we are often aware of the stages we go through to reach the propositional attitude that is the end product of the thinking process. Or, if the thought process is so quick and routine that we don't have such awareness, we can retrace our steps by deliberately seeking to justify our conclusion and so gain awareness of our thought process retrospectively. But, with respect to perception, things don't seem to be like that. I open my eyes, orient them to the world and have a perceptual experience without having any awareness of executing an extended process the earlier elements of which justify the later elements. So there would appear to be a substantial contrast between thinking and perception once again. But on second thoughts the contrast might not be so great. Just because we are not aware or conscious of any stages of inference in the process of perception, it does not follow that there are no such stages once we countenance the possibility that much might go on in the mind that is unconscious. Indeed, as we shall see in later chapters, in the 1950s and 1960s psychologists began to hypothesize that perception does involve unconscious inference.[2] If one takes this idea seriously, then one has grounds for including perception in the domain of cognition. Moreover, similar points can be made concerning other mental states and processes. Suppose I am in a café having a cappuccino. I reach out to grasp the cup, raise it to my lips and take a sip. This is a routine everyday event, and when it takes place I am not aware of my having done much in the way of thinking. But if we try to make sense of how we carry out such actions it is clear that an awful lot of complex internal processing must be involved. Once again, in the 1950s and 1960s the idea developed that our actions are driven by thought-like inferential processing that draws upon and coordinates constantly changing information about the state of the external world and one's own body coming from a range of distinct sources. So we might have grounds for regarding action and motor control as belonging to the domain of cognition.

A third point is that, even if there is a significant difference between thinking and other aspects of the mind such as perception and action, the latter must interface with the former. Consider perception first. Perceptual experiences might not be propositional attitudes or the upshot of thought processes, but they do play a key role in determining our propositional attitudes. For we routinely form beliefs and update our stock of beliefs on the basis of our perceptual experiences. Were this not the case, then perception would be of little use to us, as its value resides in its ability to provide us with knowledge about the world that we can utilize in deciding how to act so as to satisfy our needs and desires. For example, suppose you are in the supermarket looking for a tomato and while in the fruit and veg section looking at a tomato you have a visual experience as of a spherical, red object. If this experience is going to help you to satisfy your desire for a tomato, you must take it at face value and come to believe that there is a spherical, red thing before you, and you must further infer the belief that that spherical red thing is a tomato. All this clearly requires the output of perception to be taken up by thought processes. A parallel point can be made about motor control and action. Even if the movements that we make when we act are not directly driven by thought processes, those movements must be routinely and systematically motivated by our propositional attitudes. Were this not the case then our propositional attitudes would be robbed of their central function of enabling us to act so as to satisfy our desires. For example, there wouldn't be much benefit in desiring an elderflower pressé and (correctly) believing that there was a bottle of elderflower pressé in the fridge if that belief and desire pair were incapable of making an impact upon how one acted. In short, then, it is a fundamental property of thinking that it interfaces with both perception and action. This suggests that one should resist drawing a fundamental divide between perception and action, on the one hand, and thinking, on the other, and view only the latter as belonging to the domain of cognition.

I began this section by asking what cognition is and put forward the suggestion that cognition is thinking understood in commonsense terms as a process involving inferring propositional attitudes from other propositional attitudes. However, I argued that, although such thought processes are clear-cut cases of cognition, there are other processes that take place in the mind, namely those involved in perception and action and motor control, whose similarities and relationship to thinking are such as to imply that they count as cognitive processes. Thus, if cognitive science is the scientific study of cognition, it should be concerned as much with perception and action as it is with central cases of thinking.

3 Science and cognition

In the previous section I gave a preliminary account of cognition. That account implies that cognitive science is the scientific study of such phenomena as thinking, perception and action. Thus, the concern of the cognitive scientist is to explain how we humans are able to think, perceive and act, to uncover precisely what goes on within our minds when we exercise such cognitive capacities. If successful, cognitive science will reveal the core properties of humans that enable us to cognize and thus differentiate us from all the inhabitants of the universe that are not capable of cognition. Put this way, it is clear that cognitive science rests on the assumptions both that cognition is the kind of phenomenon that is amenable to scientific investigation and that we humans are cut out to execute such a scientific investigation. However, such assumptions are not universally held within the philosophical community. In this section I will describe some of the most prominent reasons for scepticism about the viability of cognitive science.

A first reason for scepticism relates to a metaphysical view about the mind most associated with Descartes (1985) and is generally known as substance dualism. According to Descartes, the human individual is a two-component system consisting of a body and a mind. The body is an inhabitant of the physical world and so, reflecting the nature of the physical world in general, is a mechanical system whose behaviour is governed by laws of nature that can be stated in mathematical terms. Moreover, the body is essentially spatially extended; that is, it has a spatial location and takes up physical space. In virtue of all this, the human body is precisely the kind of system whose nature and workings can be investigated from the scientific perspective. The mind, on the other hand, is essentially a thinking thing. It does not inhabit the physical world and, in virtue of having free will, it is not a mechanical system governed by laws of nature. Nevertheless, the mind is harnessed to the physical body and engages in a systematic two-way process of causal interaction with it. For example, the body can affect the mind, as when the physical stimulation of the sense organs causes a perceptual experience, and the mind can affect the body, as when a decision to execute a particular action causes the body to move. For Descartes, in virtue of being a fundamentally different kind of thing than the body, the mind, and therefore cognition, is not amenable to scientific study. If Descartes is correct on this point, cognitive science is not a viable enterprise.

How should we respond to this Cartesian line of thought? Descartes produced a number of arguments for dualism, but the general view of the philosophical community is that none of these are successful. Moreover, it is widely held that dualism cannot make sense of the existence of causal interaction between the mind and the body: if mental phenomena reside outside of the physical domain, how could they cause or be caused by physical phenomena taking place within the body? Indeed, as we shall see, most mainstream cognitive science is underwritten by a commitment to the anti-dualist view that the mind is in some substantial respect the brain, so the study of cognition is the study of the workings of a physical system.

A second source of scepticism regarding the viability of cognitive science accepts that cognition is the kind of phenomenon that could in principle be investigated scientifically but questions whether we humans are up to the task of conducting such an investigation. One way of characterizing this line of thought is in terms of Noam Chomsky's (2000) distinction between problems and mysteries. A problem is a question that is difficult to answer but which we have some hope of answering. A mystery is a question which we have no hope of answering because it is beyond our cognitive powers to do so. Which questions are problems and which mysteries can be species relative. For example, questions we humans find easy to answer might be utterly mysterious to a rat. We humans are bound to have cognitive limitations which make some questions that we can frame mysterious to us. One popular suggestion as to what might be a mystery for us is phenomenal consciousness. Colin McGinn (1989b) argues that phenomenal consciousness, the 'what it is like' aspect of our experiences (Nagel, 1974), is a physical phenomenon but that we are incapable of explaining how physical phenomena can generate consciousness.[3] Now I haven't characterized phenomenal consciousness as belonging to the cognitive domain, but it wouldn't be outlandish to argue that questions about human cognition are mysteries for us. For, in attempting to answer such questions, we are attempting to use our cognitive powers to understand our cognitive powers.

I don't think we should expect to be able to produce a knock-down response to this kind of scepticism. Perhaps the most sensible response would be to argue that we should evaluate the power of the objection in light of the success of our ongoing attempts to explain cognition. Thus, the power of the scepticism would be undermined to the extent that cognitive scientists produced theories and explanations that were productive and successful when judged by the criteria operative in other scientific domains.

A third source of scepticism concerning the viability of cognitive science relates to my characterization of the discipline in terms of our commonsense understanding of ourselves. For, one might argue, the phenomena that seem significant from a commonsense perspective might not be tractable from a scientific perspective, in that they are too complex and messy in being the products of many distinct but interacting factors, each of which belong to different scientific domains. Oddly enough, this idea can be found in the work of Chomsky, who, to my mind, has made one of the most valuable contributions to cognitive science.

Suppose that I utter a particular sentence. To understand me you will have to cognize what individual words I produced and how they were put together to make the sentence in question (that is, the grammatical or syntactic structure of the sentence). This is no mean feat. For one thing, there is no neat correspondence between words and sounds. This is because distinct utterances of one and the same word can be very different at the sonic level; think of how different the word 'dog' sounds when uttered by a small child, a young woman and an elderly man. And, with respect to syntactic structure, that is not an immediately perceivable property of our utterances, as is indicated by the phenomenon of structural ambiguity. For example, the sentence 'he chased the dog with the stick' could have a structure such that it means that the dog had the stick or one such that it means that the man had the stick. But suppose that you overcome such problems and work out what individual words I uttered and how I structured them in the sentence. If you know the meaning of the individual words of the sentence, you will be able to further cognize the literal meaning of my utterance, assuming that the literal meaning of a sentence is exhaustively determined by the meaning of its component words and its syntactic structure.

However, we normally understand more than the literal meaning of the sentences we hear, and successful communication depends upon this. For we appreciate the communicative intentions of the speaker (Grice, 1975). For example, suppose that you come into my office and I say 'There's a nice fresh breeze coming through the door.' I might be using that sentence with the literal meaning that it has to say any one of several things. I might be aiming to state a fact that does not require you to do anything. Or I might be aiming to point out that you have left the door open and be requesting that you close the door. If communication between us is to be successful, you must appreciate what I am aiming to do in producing the sentence – what my communicative intentions are. But if you are to do this you need to draw upon an appreciation of a potentially wide and disparate range of information. In this case such information might relate to my facial expression, my tone of voice, the temperature conditions in the room, your history of interactions with me, the likelihood of my desiring privacy for our meeting, the conventions governing meetings in my institution, and so on. Now much of this potentially relevant information lies outside of the body of linguistic knowledge that one might think we need to be capable of producing syntactically well-formed sentences and appreciating the literal meaning of the sentences we encounter. Chomsky's point is that, because understanding communicative intentions involves appreciating such a potentially wide and disparate range of information from outside the linguistic domain, it is going to be impossible for us to explain how we understand communicative intentions.

A related point is that sciences carve the world into domains of enquiry within which the scientist has some hope of making progress. But in the real world such domains often interact with one another, so that phenomena that are salient from a commonsense perspective are often massive interaction effects from the scientific perspective. Consequently, explaining such phenomena would require us to engage in 'the science of everything' – something that is not possible for us.

Chomsky's own engagement in cognitive science suggests a response to this sceptical line of thought that involves conceding the point that lies at its core. Our commonsense conception of ourselves as cognizers motivates our engagement in cognitive science and provides the discipline with a starting point of questions to address. But it is an open question as to which of those questions remain on the agenda and what questions are to replace those that are abandoned. How these issues are to be settled is something that cannot be determined in advance of our actual engagement in cognitive science. This situation with respect to cognitive science is typical of science in general, so it motivates scepticism concerning the viability of cognitive science only if it motivates scepticism about the viability of science in general.

4 Science

The upshot of the discussion of the previous section is that it would be somewhat hasty to conclude that cognitive science is not viable. But this does raise a further question: what characteristics should we expect cognitive science to have in virtue of being a science? In this section I will address this question.

Science is one of the great triumphs of Western civilization and has provided us with a systematic body of knowledge of the workings of the natural world. I am very wary of attempting to provide a general account of the nature of science. Rather, I will describe some prominent features of science that will prove to be very relevant when understanding the core commitments of cognitive science and when addressing philosophical questions about cognitive science.

First, most scientists assume that happenings in that portion of the natural world with which they are concerned are not entirely random and irregular; rather, they are governed by laws. Thus, one of the main goals of science is to discover the laws that govern the workings of the natural world. Here are some examples of such laws of nature: freely falling bodies accelerate at a uniform rate (Galileo); planets have an elliptical orbit (Kepler); the volume occupied by a gas is inversely proportional to the pressure on it (Boyle); the strength of the gravitational attraction between two bodies depends on the product of their masses and is inversely proportional to the square of the distance between them (Newton). Note that some laws (particularly those operating at the quantum level) are probabilistic rather than deterministic. That is, they are of the form that, if x is the case, then, with probability P (where P is less than 1), y will be the case. And some laws are *ceteris paribus* (all else equal) laws. Such laws have the form that, if x occurs, then y will occur, all else equal. Hence, the claim that such a law holds is not necessarily undermined by the observation that an x has occurred without being followed by a y. For, all else might not have been equal in such a case. Jerry Fodor (1987) has provided a well-known example of a *ceteris paribus* law from geology: a meandering river erodes its outer bank, *ceteris paribus*. A case where a meandering river might fail to erode its outer bank would be one where the bank had been concreted over; in such a case all else wouldn't be equal. Given the general role of laws in science, we should seriously entertain the possibility that the cognition is law governed and that one of the main tasks of cognitive science is to uncover such laws.

Second, another of the major goals of science is to explain features of the natural world. This involves not merely describing the natural world but saying why it is the way it is and how things work in the natural world. Laws are relevant here as they have an important role in explaining natural events. For example, Galileo's law plays a key role in explaining why a cannonball and a marble when dropped from a particular height at a given time will hit the ground at the same point in time, in violation of our commonsense expectations that the cannonball, in virtue of being heavier, would hit the ground first. Indeed, explaining why a particular law holds typically involves appealing to more basic laws. For example, Newton appealed to his laws of motion and the inverse square law in order to explain Kepler's law of planetary motion. My initial characterization of cognitive science highlighted the role of explanation, as I said that cognitive science is concerned with explaining our cognitive abilities.

Third, science is an empirical discipline in that observation and experiment play a central role in the scientific project. This contrasts science with any purely *a priori* or armchair activity. One prominent account of the precise way in which science is empirical was developed by Karl Popper (1959). Popper was concerned with the so-called demarcation problem, with specifying precisely what distinguished science from non-science. He argued that it was a mark of a scientific theory that it was falsifiable – that is, a scientific theory has implications as to what observable phenomena will take place in particular circumstances such that there are, in principle, observations we could make that would definitively show the theory to be false. Thus, the scientist proceeds by a two-stage process of conjecture and refutation. This involves producing a conjecture or hypothesis in order to explain some target phenomenon. The observational implications of the conjecture are then worked out and an experiment conducted in which the relevant observations are made. If the observations are inconsistent with the observational implications of the conjecture, then the conjecture is rejected as false and the scientist goes back to the drawing board. If the observations do not falsify the conjecture, then the scientist conducts further experiments in order to test and falsify the conjecture.

Fourth, in the course of explanation, scientists often postulate theoretical entities. These are entities that are not observed but are invoked to explain phenomena that are observable. For example, suppose a biologist wanted to explain why organisms generally bear many similarities to their parents but have some differences from them. The explanation proffered by the biologist will appeal to genes. The key point is that the genetic explanation[4] is not produced on the basis of observing genes and their activities. Rather, the genes are postulated in order to make sense of the otherwise mysterious observed phenomena. Other prominent examples of theoretical entities are atoms, quarks and photons. Given the role of theoretical entities in science, we should expect cognitive scientists to postulate theoretical entities in theorizing about human cognition.

Fifth, although science is empirical in nature, the role of observation in science is not as straightforward as implied by Popper's work as described above. W. V. O. Quine's 'The Two Dogmas of Empiricism' – first published in 1951 – is probably the most important philosophical article written in English since the Second World War. In this paper Quine seeks to undermine two central theses of the empiricism of his day. The second of these theses is that the meaning of a sentence is its method of verification – that is, the empirical means of determining its truth-value. Thus, any meaningful sentence can be translated without loss into a sentence about experience or observations (this doctrine is known as reductionism). In rejecting verificationism, Quine champions a position that has become known as the Duhem–Quine thesis.[5] It has this name because Quine's basic idea was anticipated by the French physicist and philosopher of science Pierre Duhem (1954), in his book *The Aim and Structure of Physical Theory*.

Suppose that one wants to determine by empirical means whether a particular (declarative) sentence is true or false. One will then have to make relevant observations using one's senses. It is natural to think that, with respect to any true sentence, it is in principle possible to establish that it is true by making relevant observations; and that, with respect to any false sentence, it is in principle possible to establish that it is false by making relevant observations. Applied specifically to science, the idea would be that, with respect to any sentence expressing a scientific theory, it is in principle possible to establish whether it is true or false by observational means. Quine's basic point is that this is not the case because it is not possible to verify a sentence or theory in isolation. This is because individual sentences or theories do not by themselves have the kind of observational implications that enable them to be verified. Rather, it is only collections of sentences or theories that have such implications. To see this, consider a topical example. Attention Deficit Hyperactivity Disorder (ADHD) is a condition attributed to many children in the Western world today. The condition manifests itself in a child's inability to concentrate on schoolroom tasks and a tendency to be disruptive. What causes ADHD and why is it so prevalent today? One theory is that it is the result of a deficiency in Omega 3. How are we to verify this theory or the sentence that expresses it? Here is an answer. Give children with the condition an Omega 3 supplement and observe their subsequent behaviour. If the children are observed to undergo an improvement in their powers of attention and concentration, then that tells in favour of the theory, suggesting that the theory may well be true or is a serious candidate for truth. If, however, the children are observed to undergo no such improvement, then that tells against the theory, suggesting that it is (probably) false. The problem is that attempting to verify the theory/sentence in this way involves making a whole load of assumptions that are independent of the theory/sentence. For example, the assumption that the supplement contains Omega 3 in a form that can be readily absorbed by the body; the assumption that if the condition is due to a deficiency in Omega 3 then it can be remedied by taking a supplement; assumptions about how concentration/attention powers manifest themselves in observable behaviour; and so on. Making such assumptions will involve committing oneself to the truth of substantial scientific theories. If one doesn't make these assumptions, then one will regard the above way of verifying the theory as illegitimate. If one makes alternative assumptions, then one will be committed to adopting a different means of verifying the theory. And if one makes no supplementary assumptions at all, then one will have no idea how to verify the theory, as the theory in itself will not tell one how to go about verifying it.

Now suppose that one does make the above described assumptions and that no improvement in the children is observed. Does this tell against the theory? According to Quine, the answer to this question is negative. For, in order to deal with the recalcitrant observational data, one can hold onto the theory but reject some of the associated assumptions. Thus Quine says, 'any statement can be held true come what may, if we make enough adjustments elsewhere in the system' (1951: 43). All this is captured in the Duhem–Quine thesis, according to which:

any theory can be held onto in the face of recalcitrant observational data by making suitable adjustments elsewhere in one's system of commitments.

The Duhem–Quine thesis is widely held by philosophers of science. With respect to cognitive science, an implication of the thesis is that it may well be difficult to adjudicate between different theories about our cognitive lives, as their advocates disagree on the significance of the empirical data they gather.

A sixth feature of science concerns its relationship to commonsense. If a human individual is going to prosper in a challenging and potentially dangerous world, they will need to have a battery of concepts in order to categorize the phenomena with which they interact. And they will need to employ those concepts to form knowledge or beliefs concerning how such phenomena behave. Armed with such knowledge, they will then be able to predict and explain worldly events and so enhance their survival prospects. When philosophers and cognitive scientists talk about commonsense, they are talking about the relatively untutored and unsophisticated conceptual schemes and associated belief/knowledge systems that ordinary people routinely employ in their everyday lives. Commonsense is therefore distinguished from mature and sophisticated scientific theories. However, that is not to say that commonsense bears none of the characteristics of science. Questions about commonsense are prominent within cognitive science, and a widely held view represents it as bearing some of the characteristics of mature science. First, it has several components each of which relate to a distinct subject matter, so that commonsense is underwritten by the assumption that there are different kinds of phenomena in the world that behave in their own distinctive way. In other words, commonsense, like science, has its component disciplines.[6] Moreover, some of them correspond to prominent scientific disciplines. For example, there is commonsense physics that is concerned with inanimate physical objects, commonsense psychology that is concerned with minded agents, particularly humans, and commonsense biology that is concerned with biological entities such as animals. Second, commonsense concepts can be quite abstract in the respect that they group together phenomena that differ widely in terms of their perceivable properties, such as their shape, size and colour, and in the respect that they refer to the unobservable. For example, commonsense physics employs a quite general concept of a physical object that is utilized in such beliefs that unsupported physical objects fall to the ground. And commonsense biology employs the concept of a hidden essence, in that we think of creatures as having characteristics we cannot directly perceive that determine the kind of thing they are and which are causally responsible for the non-essential surface characteristics that we can perceive (Keil, 1989).

One quite natural view of the relationship between science and commonsense is that the former is born of the latter. Science is a cultural phenomenon which in an organized and rigorous form has a relatively short history of approximately 500 years and so is very much pre-dated by commonsense. Commonsense constituted the starting point for science, in that it provided it with its core questions and conceptual scheme. For example, if commonsense physics assumes that physical objects causally interact with one another and are governed by generalizations that advert to their physical properties, then those assumptions generate questions for the physicist – questions as to the underlying nature of these physical properties (for example, what is heat?), the underpinnings of the generalizations (for example, why do unsupported bodies fall?) and the identity of any further physical properties and generalizations not currently recognized by commonsense. However, even if science is born of commonsense, it doesn't follow that in its mature form a science shares many of the characteristics of its parent. Science is a self-conscious research endeavour that is driven by a relentless search for progress and the truth in a way that commonsense is not. As a result, science often exposes the limitations of commonsense: the parochial and inadequate nature of its conceptual scheme, the falsity of its assumptions, the limits of its explanatory powers and ambitions (Churchland, 1979). To see this, just think of how relativity theory and quantum mechanics have left commonsense physics behind. In short, then, individual sciences have a basis in commonsense but, in their mature form, have often moved a considerable distance from their parent.

In this section I have described a number of key characteristics of science in general. In virtue of its status as a science, we should expect cognitive science to share these characteristics. Thus we should expect cognitive science to (i) seek to uncover laws governing the workings of the cognitive mind; (ii) utilize such laws to explain cognitive phenomena; (iii) appeal to unobservable entities in its laws and explanations; (iv) utilize observation and experimentation; (v) confirm theories in a Duhemian–Quinean manner; and (vi) be born of yet to have moved beyond commonsense psychology.

5 The birth of cognitive science

So far I have been talking quite generally about cognitive science, but it is now time to be a little more specific about its origins and core commitments. Attempts to study cognition from a scientific perspective are hardly new. For example, David Hume ([1738] 1978) described the project executed in his work *A Treatise of Human Nature* as being an attempt to do for the mind what Newton had done for the external physical world. Indeed, some of the concepts and ideas employed by Hume have their analogues in contemporary cognitive science. Nevertheless, it is generally assumed that cognitive science came into existence as a discipline with a distinct identity only in the late 1950s and early 1960s with an intellectual turn known as the cognitive revolution.

The cognitive revolution was a revolt against a movement in psychology that dominated that discipline, at least in the English-speaking world, throughout the first half of the twentieth century. The movement in question was behaviourism, which had J. B. Watson (1913) as one of its pioneers and B. F. Skinner (1953) as its most prominent advocate in its maturity. Psychology as the study of mind and behaviour had become established as a discipline studied in universities in the late nineteenth century.[7] During its infancy, psychology witnessed several different approaches, chief among them being introspectionism and psychoanalysis. Introspection is the process of observing one's own mental states and so involves the mind turning its gaze in on itself. Hence, introspectionism is an approach in psychology that relies upon an individual's testimony as to what they have introspected. For a psychoanalyst such as Freud, the limitation of introspection is that much of what goes on within the mind is unconscious. Hence, psychology should involve postulating hidden phenomena from a less subjective third-person perspective.

For the behaviourist, neither introspectionism nor psychoanalysis constitutes an acceptable means of conducting psychology. The problem with them is their failure to be scientific, as they focus their concern on phenomena that are either subjective or hidden, with the upshot that their theories are unconstrained by observable evidence. In order to be scientific, the behaviourists thought, psychology must focus on observable phenomena and so turn its attention away from inner, mental phenomena.

The core concepts employed by the behaviourists were those of stimulus, response and reinforcement, and they can be understood by considering the work of B. F. Skinner. For many years Skinner worked with animals such as rats and pigeons where his central concern was to control and predict their behaviour. In a classic Skinnerian scenario, a rat would be placed in a box containing a bar that can be pressed. In this context, a stimulus is some feature of the environment that can impinge upon the rat. When the rat is first placed in the box it is not disposed to behave in any particular way in response to the stimulus constituted by the bar. Suppose that, behaving randomly, it presses the bar, and further suppose that a food pellet is dispersed as a result of this behaviour. The dispersal of the food pellet will act as a further stimulus that will serve to reinforce the behaviour of bar-pressing in response to being confronted by the bar in the box; that is, it will increase the probability or frequency of the rat making such a response to the stimulus in future interactions. In this way the rat's behaviour can be conditioned or trained. If the dispersal of the food pellet were made conditional on the presence of some other stimulus such as the flashing of a light, then the rat can be conditioned to press the bar only when the light flashes. And if the dispersal of the food pellet is further made conditional on the bar being pressed with some specific force, then the rat can be conditioned to press the bar with that specific force when the light flashes. In short, the rat's behaviour at any given point will be a product of the current stimulus and its history of reinforcement and so can be controlled by reinforcement and predicted on the basis of knowledge of the stimulus and history of reinforcement.

Skinner's success in training animals led him to argue that human behaviour is not fundamentally distinct from that of rats and pigeons, in the sense that it too is a product of the current stimulus and history of reinforcement. Thus, in his book *Verbal Behaviour*, Skinner (1957) argued that language learning is a matter of reinforcement; so, for example, learning the English word 'horse' is a matter of being reinforced to behave by uttering 'horse' in response to horses.

In 1959 Noam Chomsky published a blistering review of *Verbal Behaviour* which is often characterized as a seminal event in the history of cognitive science, in that it revealed the bankruptcy of the behaviourist approach and the need for an alternative that focused on our inner mental workings. Chomsky's review is lengthy, subtle and intricate and gestures towards several of the ideas for which he was to become famous. I can hardly do justice to it in this context and, accordingly, will restrict myself to highlighting just a few of his points. First, Chomsky argues that Skinner's success in studying animals doesn't carry over to the study of humans. If we understand terms such as 'stimulus', 'response' and 'reinforcement' literally, then those terms don't apply to humans, and if we understand them metaphorically they just turn out to be mentalist terms in disguise. Second, when it comes to animals, sometimes their behaviour is learned but sometimes it is a product of a genetically controlled process of maturation. So, in advance of studying the relative contributions of the kind of factors highlighted by Skinner, on the one hand, and the organism's contribution to learning, on the other, it is plain dogmatic to adopt the behaviourist approach and ignore the internal workings of the organism under study. Third, with respect to language learning, it is clear that Skinner offers little insight. Children typically learn language quickly and reliably without being provided with much in the way of organized reinforcement. Hence, if we are to understand language acquisition, we must uncover the internal mechanisms that facilitate learning.

The general moral of Chomsky's review is that behaviourism must be replaced by a serious study of the contribution made by the internal workings of the human organism to behaviour and learning. This call did not fall on deaf ears and proved to be a huge factor in motivating the emergence of cognitive science. But it could only motivate the emergence of cognitive science because of the availability of a number of ideas that could be brought together to provide cognitive science with a body of core assumptions and methodological principles. Those ideas came from different disciplines: some were of considerable vintage and some had a presence in more than one discipline; it was their coalescence that gave birth to cognitive science as a recognizable coherent discipline. I will now describe the ideas in question and how they cohere so as to ground cognitive science during its infancy.

A first idea was widely endorsed by philosophers during the modern period of the seventeenth and eighteenth centuries. This is the idea that the mind is populated by representations.[8] A representation is a symbol residing in the mind, and mental representations are involved whenever a person is in a particular mental state (for example, has a belief or a perceptual experience) or executes a mental process (for example, thinks or recollects a past event). Just as external representations such as spoken or written words, pictures or maps have meaning or content, so too do (mental) representations, and the content of a mental state is inherited from the content of the representation involved in having it. One prominent question about mental representations is why they have the content that they have. One answer[9] is that mental representations are imagistic in nature and their meaning is grounded in resemblance, so that, for example, the representation involved in having thoughts about horses means *horse* in virtue of resembling a horse.

The idea that representations are involved in perception and cognition was endorsed in the nineteenth century by many psychologists who were influenced by philosophers of the modern period. For example, Wundt was influenced by Kant and William James was heavily influenced by Hume.

However, the idea that representations are involved in cognition became unfashionable in the first half of the twentieth century in both philosophy and psychology. Of course the behaviourists would have no truck with representations, given their inner, mental nature. Moreover, they had a specific problem with them, namely, that the postulation of representations inadvertently involves a postulation of homunculi, little intelligent agents residing in the mind. This kind of objection was independently developed by philosophers such as Wittgenstein (1953) and Gilbert Ryle (1949),[10] from where it takes the following. External representations do not mean what they mean in and of themselves but rather mean what they mean in virtue of how we use and understand them. As contemporary philosophers put it, their meaning is derived rather than original (Searle, 1992). But such use and understanding is a mental phenomenon that exhibits intelligence. The meaning of a mental representation would similarly be a product of its intelligent use and understanding, so that to postulate such a representation would be inadvertently to postulate an intelligent agent in the mind (a homunculus) to provide it with its meaning. But that would raise the question of the basis of the intelligence of the homunculus. If we postulate further representations in its head, then an infinite regress looms, and if we leave its intelligence unaccounted for, then our attempt to understand or explain cognition counts as circular. It is no good responding to this objection by pointing out that, if the representations are images, then their meaning is grounded in resemblance, a relationship that is objective and does not require intelligence. This is because, as Wittgenstein argued, resemblance is not an objective relation at all and resides very much in the eye of the beholder. For example, a picture of an old man walking up a hill resembles the phenomenon of an old man sliding down a hill just as much as that of an old man climbing a hill. What settles the ambiguity is how the picture is used and understood. In the face of such worries, the postulation of representations fell out of favour and was to be resurrected only with the emergence of a second idea.[11]

The second idea involved in the birth of cognitive science comes from logic and mathematics and is associated particularly with Alan Turing. Turing invented a very simple, abstract computing machine that has become known as a Turing machine. He didn't actually build any models of this machine; rather, he theorized about such machines and attempted to prove logico-mathematical theses about them. What he established in this way is that any computable function can be computed by an appropriate Turing machine. The term 'function' is to be understood in the mathematical sense as a mapping of numbers onto numbers. For example, addition is a function which maps the numbers 2, 2 onto 4, the numbers 6, 3 onto 9, and so on (whereas subtraction maps those same orderings of numbers onto 0 and 3 respectively). A computable function is one that can be mechanically computed in a finite number of steps. What this means is that there is a procedure that, given as input any items in the domain, can mechanically work out in a finite number of steps what value in the range the function maps those items onto.[12] A universal Turing machine is a Turing machine that can be programmed to mimic any other Turing machine. What Turing further established is that any computable function can be computed by a universal Turing machine.

The idea that Turing's work suggests is that the mind is a Turing machine or ensemble of such machines, so that cognition is a form of computation that involves the mechanical manipulation of symbols. To see how Turing's work suggests this idea we need to know a little bit more about Turing machines. A Turing machine consists of an infinitely long tape divided into squares. Such squares can be filled with a symbol such as '0' or '1' or remain blank. The machine also consists of a read–write head which scans squares on the tape one at a time. When it scans a square, the read–write head is able to detect whether it is blank or filled with a '0' or a '1'. How it responds to what it detects will depend upon its internal state, for the read–write head is capable of being in a number of distinct internal states. Whatever internal state it is in, the response will have a common complex form that has a number of elements: it will involve (i) either leaving the symbol unchanged or deleting it and replacing it with some other symbol; (ii) moving one square to the left, one square to the right or halting; (iii) remaining in the same state or changing into some other state. A Turing machine's machine table is a specification of how it responds to any possible symbol for each of the internal states it is capable of being in. This table can be thought of as a series of rules or instructions that constitute the machine's program and which is hard-wired into the machine. A machine table is such that, when a Turing machine is fed a section of tape, it will go through a procedure of scanning the squares of that section, changing some of the symbols that it so detects and changing from one internal state to another as it does so, until finally it comes to a halt. What the machine will have done is taken input in the form of a section of tape with symbols printed on it and produced as output a section of tape with symbols printed upon it. These strings of symbols can be regarded as representations of numbers in the binary notation. Suppose a particular machine has a table such that, whenever it is fed a section of tape with n distinct numbers printed on it in binary notation, it responds by producing a section of tape with the number that is their sum printed on it in binary notation. Then the machine will be an adding machine that computes the addition function. In short, then, what Turing machines do is take symbols as input and produce symbols as output and, in so doing, compute mathematical functions.

What Turing's work suggested to philosophers and psychologists interested in cognition was a way of resurrecting the old idea that cognition involved representations without running into the homunculus problem. Solving mathematical problems is precisely the kind of thing we do when we cognize. As Turing machines solve mathematical problems by means of manipulating symbols without relying on intelligence or insight, they are hardly homunculi. So, if one thinks of the mind as a computer akin to a Turing machine, then one can have representations without homunculi.

A third idea that contributed to the birth of cognitive science relates to the metaphysics of the mind. Philosophers have long been concerned with the metaphysical question of the nature of the mind and its relationship to the physical. Earlier I described Descartes' substance dualist view and characterized it as problematic for anyone who wants to study the mind and cognition from a scientific perspective. During the nineteenth and twentieth centuries dualism waned in popularity among philosophers and materialism or physicalism became the orthodox view. But to say that the mind is physical or material isn't in itself to present a fully fledged theory. However, appropriate theories that could underpin cognitive science gradually emerged.

Even Descartes thought that there was a particular intimate relationship between the mind and the brain, in that the interface between the two systems lay at the pineal gland in the brain. Developments in neurophysiology in the nineteenth century suggested an even closer relationship, especially with the discovery – by the likes of Broca and Wernicke – that distinct components of the brain were directly associated with particular cognitive capacities. In short, the more we came to understand about the workings of the brain, the clearer it became that there was a direct and intimate relationship between the mind and the brain. It is a short step from this to the conclusion that in some substantial respect the mind is the brain, for, one might ask, how could there be such an intimate connection were the mind not the brain? It is no good answering this question by postulating a causal relationship that implies a distinction between the two, as that raises the problem of how two fundamentally different kinds of thing (the physical brain and the immaterial mind) could causally interact.

However, the theory that the mind is the brain can take forms which are problematic for cognitive science. To see this, consider a popular mid-twentieth-century view known variously as type-type physicalism, the mind–brain identity theory and central state materialism.[13] The basic idea is that types of mental state are identical to types of brain state, or, alternatively, mental properties are identical to neural properties. Here the term 'identical to' means 'one and the same as'. These mental–neural identity relationships were conceived as being akin to such familiar identity relationships as those holding between water and H_2O and heat and mean kinetic energy. Thus, with respect to any particular type of mental state M (e.g., pain, the belief that coffee contains caffeine), answers to such questions as 'What is M?', 'What is it to be in M?' and 'What do all creatures that are in M have in common in virtue of which they are in M?' will appeal to a particular type of neural state or property. The type identity theorists held that these identity relationships needed to be discovered by scientists, but as an example they often claimed that pain is C-fibre firing.

The type-type identity theory implies that, in essence, mental states are neural states. Thus, if I describe one of my internal states as a desire for a cup of coffee, I haven't described that state in terms of what, from the scientific perspective, are its core properties. Rather, I have described it from a commonsense, pre-scientific perspective. That I do this is a product of my own ignorance, so that, when science corrects that ignorance and finds a means of understanding the brain from the neural perspective, talk of beliefs, desires, and the mental and cognitive in general should be abandoned and replaced by purely neural talk. This implies that we should be aiming for a neuroscience unsullied by mental and cognitive talk rather than a cognitive science that inherits our commonsense perspective of ourselves as cognizers.

In short, then, cognitive science needed a respectable metaphysical view about the mind that, though anti-dualist, contrasted with the type-type

identity theory. Such a theory developed in the early 1960s out of a recognition of a key failing of the type-type identity theory. This failing was
identified by Hilary Putnam (1967), who accused the type-type identity
theory of being chauvinistic in that it denied mentality of systems (be they
earthly creatures, extraterrestrials or inorganic machines) physically unlike
humans. For the theory implies that only systems with brains similar to
ours are capable of sharing any aspect of our mental life. Putnam pointed
out that this implication is highly implausible, as we are confident that
many animals share aspects of our mental lives (for example, feel pain and
have thoughts) even though they have central nervous systems that are
very different from ours.

In the light of this failing, Putnam sought to develop a non-chauvinistic
theory, and what he came up with was functionalism. According to functionalism, mental types are functional types. Here the term 'functional' is
to be understood in causal rather than teleological terms. What the particular instances (the tokens) of any distinct mental type have in common
in virtue of which they belong to that type is the functional or causal role
that they play. It would perhaps help to consider an example of a mental
state, namely pain, that isn't a cognitive state in order to illustrate the
functionalist approach. Pain plays a distinctive causal role in our internal
economies, a role that is specified by generalizations such as these: pain is
caused by bodily damage or certain kinds of nerve stimulation; pain causes
worry; pain causes moaning, groaning and crying; and so on. According
to the functionalist, occupying this causal role is not a contingent feature
of pain; rather, it is part of its essence. Thus to be in pain just is to instantiate an internal state that occupies the relevant causal role (which stands
in the appropriate causal relations to inputs, outputs and other mental
states). Generalized, the view is that each distinct type of mental state,
including cognitive states, plays a particular distinctive causal role that is
central to its identity and that to instantiate a state of any given mental type
is to instantiate an internal state that plays the appropriate role in one's
internal economy. Most functionalists are physicalists in that they regard
minded systems as being complex physical systems and hold that the token
states that occupy these roles are internal physical states of such systems.
However, mental states are multiply realizable at the physical level. That
is, for any given type of mental state M, the physical state that occupies the
M role in one system may vary considerably from that which occupies the
M role in another system. For example, the pain role may be occupied by
C-fibre firing in humans, O-fibre firing in octopuses, and yet another state
in a Martian with a silicone-based chemistry. Thus, the chauvinism prob-
lem that dogged the type identity theory is avoided.

An important feature of the causal roles of mental states is that they
involve interacting with other mental states in distinctive ways; for example,

pain causes worry, a belief that one's body is in danger of being damaged, a desire for the pain to stop, and so on. Moreover, how a mental state manifests itself in behaviour will depend upon what other mental states one is in due to the fact that mental states cause behaviour in concert with other mental states (again, this is central to their causal role). This implies that one can't characterize a mental state without referring to other mental states.

In actual fact, Putnam appealed to Turing machines in developing his own version of functionalism. Particular Turing machines are defined by their machine table, which specifies the relations between possible symbolic inputs, symbolic outputs and internal states. Turing machines are multiply realizable at the physical level in that it is possible to build a given type of Turing machine out of different materials (all such physically divergent machines will satisfy the same machine table). With respect to the states of the machine, they are implicitly characterized in terms of their relations to inputs, outputs and one another and not in terms of the physical form they take in any particular concrete machine. Putnam argued that mental states are Turing machine states. In developing functionalism along these lines, Putnam made clear its affinity with the idea that cognition involves computation and so revealed its value as the metaphysical theory of the mind needed by cognitive science.

A fourth idea that played a key role in the birth of cognitive science relates to consciousness and self-knowledge. Descartes is associated, in addition to dualism, with the view that we have a thoroughgoing consciousness or knowledge of the contents of our own minds. On this view, if I am thinking a particular thought, I cannot but know that I am thinking that thought. From the commonsense perspective this view is very appealing, as we do seem to have a direct and immediate means of access to the contents of our own minds that we do not have to the contents of the minds of our fellows, which makes us authoritative about what we think.

This idea of the extent of our self-knowledge was, in effect, attacked by Freud in his postulation of the unconscious. For Freud, mechanisms of repression ensure that many of us have beliefs and desires that, because of their potentially repellent or unsettling nature, are pushed into parts of the mind to which we have no direct access. Nevertheless, such unconscious mental states can manifest themselves in behaviour in such a way that a skilled third party – such as a trained analyst – who has knowledge of the individual's history can uncover them. In short, Freud's work suggests that there is much in an individual's mind that is unconscious and that it is possible to know from a third-person perspective that an individual has a particular mental state even when they do not have such knowledge from the first-person perspective. Both Freud's work and the psychoanalytic movement that it spawned are in many ways highly controversial.

However, they have made the idea that we have unconscious mental states both familiar and popular. In fact, I would go so far as to say that Freud's work has impacted on commonsense in such a way that it is part of our contemporary commonsense vision of ourselves that some of our mental states are unconscious.

Although cognitive science and Freudian psychoanalysis are in many ways poles apart, the rise of the latter in the late nineteenth and early twentieth century popularized the concept of the unconsciousness and, in so doing, paved the way for the idea that many of the processes and states involved in cognition are unconscious. This idea is a core idea of cognitive science.

In this section I have described four ideas, namely, that the mind is inhabited by representations, that cognition involves computation, that mental states are functional states, and that much of what goes on in the mind when we cognize is unconscious. We have seen that some of these ideas very much pre-date cognitive science and come from very different sources. These ideas made cognitive science possible, and their coming together in the late 1950s and early 1960s both gave birth to cognitive science as a distinct and self-conscious discipline and provided it with its core theoretical assumptions.

6 Interdisciplinarity

Cognitive science is often described as interdisciplinary in nature, where the main contributing disciplines are usually identified as psychology (particularly cognitive and development psychology), Artificial Intelligence, neuroscience, philosophy and linguistics. Cognitive science is interdisciplinary for two reasons. First, the core ideas that came together at its birth originate from different disciplines, so that early cognitive scientists were thereby engaging with and combining ideas that have different disciplinary homes. Second, the core ideas provide a picture of cognition which implies that several traditionally distinct disciplines have major contributions to make. Cognitive psychology has a role to play as it is the branch of psychology concerned with cognition, along with those parts of developmental psychology that are concerned with how our cognitive capacities develop from birth onwards. Neuroscience has a role to play in revealing how cognitive processes are ultimately implemented in the brain and in placing constraints on higher-level theories of cognitive processing that are not directly concerned with their neural implementation. Artificial Intelligence is the project of programming computers to behave in a way that would count as exhibiting intelligence if done by a human individual. The strengths and limitations of particular AI programs can suggest hypotheses as to how we cognize and provide evidence concerning the plausibility of such hypotheses.

The role of linguistics in cognitive science is a little less direct and obvious and has to do with the work of Noam Chomsky and the approach in linguistics that he initiated.[14] We have already seen the significance of Chomsky's attack on Skinner and his demand that we concern ourselves with the contribution of the organism in studying linguistic behaviour. But he also went on to develop a view of language that implies that linguistics, in virtue of being the study of language, belongs to cognitive science. For Chomsky, language is not a social entity such as a body of social conventions or practices that exist externally to the minds of individual speakers of the language. Rather, an individual's language is an internal state of her mind. Thus, in studying language, the linguist is studying the mind. Moreover, Chomsky demands that linguists produce theories of language which explain language acquisition, so that linguistics is also concerned with cognitive development, in particular, the process by means of which a child moves from the state where she appears to have no grasp of language to that where she is a competent mature speaker.

Finally we come to the place of philosophy in cognitive science. Cognitive science is concerned with many of the issues that have been central problems throughout the history of philosophy, issues as to the place of the mind in the physical world, the nature of thinking, the relationship between mind and language, the role of learning in cognitive development, and so on. But this is not a straightforward case where philosophical speculation flourished in the absence of scientific insight only to lose its relevance at the hands of scientific advance. For, as we have seen, many of the core concepts and insights lying at the basis of cognitive science came from philosophy. Indeed, as will be made clear throughout this book, philosophy has continued to play such a role and so has an especially important role in adjudicating competing claims as to the importance of distinct bodies of data coming from different disciplines. Hence, philosophy is an important partner in the cognitive scientific enterprise.

7 Conclusion

In this chapter I have provided a general account of the nature of cognitive science and how it emerged in the late 1950s and early 1960s. I have indicated that in its early incarnations it was based on a commitment to the idea that cognition is a form of computation. In the next chapter I will examine in some detail this commitment to computationalism and how an alternative version of the computationalist vision emerged in the 1980s in the form of connectionism. I will then consider some recent challenges to this shared conception that internal representations play a central role in cognition.

2 Representation and Computation

1 Introduction

In the previous chapter we saw that, at its inception, cognitive science was committed to the idea that cognition involves the manipulation of representations by means of computation. These representations and representation-manipulating computational processes are implemented or realized in the brain. The upshot of this is that explaining a particular cognitive capacity, such as visual perception, object recognition, high-level reasoning, action planning, language development, understanding the mental states of another person, and so on, involves identifying the representations and computational processes involved whenever that capacity is exercised. But what general form do these representations and computational processes have? Within cognitive science there are two broad answers to this question, reflecting a divide between two competing approaches. The first answer is associated with an approach widely known as classical computationalism. As its name suggests, classical computationalism dominated cognitive science during the early decades of its existence.[1] The second answer is associated with an approach known as connectionism, which, though having its origins in work in the 1950s,[2] came to the fore only in the 1980s with the publication of Rumelhart and McClelland's connectionist 'bible' (Rumelhart et al., 1986; McClelland et al., 1986). Both connectionism and classical computationalism are very much alive today. However, recent years have seen the emergence of a family of approaches offering a radical alternative to both classical computationalism and connectionism, an alternative that questions the role of mental representations in cognition. In this chapter I will examine each of these three perspectives on cognition, beginning with classical computationalism.

2 Classical computationalism

A classical computer is a machine that takes structured language-like symbols as input and produces structured language-like symbols as output. This input–output profile is mediated by the application of rules which collectively constitute the program that the machine runs. The application of the symbol-manipulating rules – the running of the program – is a mechanical process in that it does not require any intelligence or creativity on the part of the machine.[3] In particular, it does not require the machine to understand or appreciate the meaning of the symbols that it manipulates. What the machine is sensitive to are the formal or syntactic properties of symbols, and the symbol-manipulating rules it applies relate to such properties. Hence, a classical computer is, to use Daniel Dennett's (1978c) memorable phrase, a 'syntactic engine'. Nevertheless, the manipulated symbols typically have semantic properties – after all, they are symbols, and it is natural to think that, in order for an item to be a symbol, it must either have some meaning or be apt to have some meaning attributed to it. Moreover, the meaning of the output symbols generated by a classical computer is typically coherently related to the meaning of the input symbols. A simple example of this would be a machine that takes as input pairs of Arabic numerals and produces as output the numeral that is the sum of the input pair. Thus, for example, the machine gives the output '5' in response to the input '2, 3', '10' to the input '3, 7', and so on. Thus, the machine computes the addition function for pairs of numbers despite the fact that it doesn't have any understanding of mathematical concepts such as numbers and addition. Therein lies the beauty of classical computers: despite the fact that they are entirely mechanical and incapable of understanding, they can be built so as to mimic the behaviour of a system that does have understanding and intelligence – for example, a system that has mathematical understanding and intelligence.

What I have said so far could do with some unpacking. First of all, what is involved in claiming that the symbols manipulated by the machine are language-like structures that have syntactic properties? Consider a natural language such as English. It has a vocabulary or lexicon consisting of finitely many words, each of which belongs to a particular syntactic category. For example, the words 'dog' and 'postman' are both nouns, 'a' and 'the' are both determiners, 'in' is a preposition, 'and' is a conjunction, 'chase' is a verb, 'angry' is an adjective, and so on. These words can be combined to create larger and more complex structures such as phrases and sentences. But they can't be combined in any old way, as English has a body of syntactic rules that determine how words can be combined. So, for example, 'a dog chased the postman' is a legitimate sentence of English, as that combination of words is permitted by the syntactic rules of English. 'The a postman dog chased', on the other hand, is not a legitimate sentence of English, as it cannot be derived by the application of the syntactic rules of English to the words of that language.

One might reasonably ask what the syntactic rules of English are. This is no easy question to answer and there is much controversy in linguistics, not least because speakers of English don't have conscious knowledge of the syntactic rules of their language. But, to get a flavour of what those rules might look like, consider some of the basic phrase structure rules postulated by linguists working in the generative grammar tradition initiated in the 1960s by Noam Chomsky (1965).[4]

1   A sentence consists of a noun phrase followed by a verb phrase.
2   A noun phrase consists of a determiner followed by a noun.
3   A verb phrase consists of a verb followed by a noun phrase.

Now we can apply these rules to the words 'the', 'a', 'postman', 'dog' and 'chased' – which are, respectively, a determiner, a determiner, a noun, a noun and a verb – to create more complex phrases and sentences. For example, one can apply rule 2 to 'the' and 'postman' to get the noun phrase 'the postman'. One can then apply rule 3 to that noun phrase and 'chased' to get the verb phrase 'chased the postman'. Rule 2 can be applied to 'a' and 'dog' to get the noun phrase 'a dog'. Finally, rule 1 can be applied to the previously generated noun phrase 'a dog' and verb phrase 'chased the postman' to get the sentence 'a dog chased the postman'.

Clearly, rules 1 to 3 can be applied to those same words to generate several other sentences, such as 'the postman chased a dog', 'a postman chased the dog', 'the postman chased the dog', and so on. But note that there is no way of getting 'the a postman dog chased'. Now if we added some more words and more syntactic rules, for example, relating to adjectives, prepositions, conjunctions, and so on, the number of legitimate phrases and sentences would mushroom.

Sentences and phrases have syntactic properties. The syntax of a phrase or sentence is a matter of the syntactic category to which each of its constituent words belong and the manner in which they are put together by application of the syntactic rules. Linguists often represent the syntax of phrases and sentences by means of tree diagrams such as shown in figure 2.1.

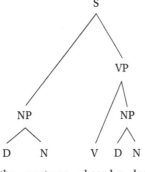

Figure 2.1

Keeping things at this simplified level of analysis doesn't bring out an important feature of languages such as English. This is that some of its syntactic rules are recursive, in that they can be applied to their own output over and over again, thereby creating ever more complex structures. I will discuss the recursivity of natural languages in some detail in chapter 5, but for now a simple example will suffice. Suppose that we add a number of words to our basic description of English given above, including the conjunctions 'and', 'or' and 'if . . . then'. In addition, suppose we add a further syntactic rule to the effect that two sentences can be combined by means of a conjunction to form a new sentence. Thus, if A is a sentence and B is a sentence, then 'A and B', 'A or B' and 'if A then B' will also be sentences. Note that if we build such sentences we can apply this rule again, to get 'if A and B then A or B', 'A or B and if A then B', and so on. And we could glue those sentences together to get yet more complex structures, and so on ad infinitum. What the presence of recursive rules implies is that a language will have infinitely many sentences, even though it has only finitely many words and syntactic rules. This feature of languages such as English is known as productivity: English is a productive language.

Thus far I haven't said anything about meaning in relation to English. Of course, the basic words of English are meaningful, and they contribute that meaning to any sentence in which they figure. A widespread assumption is that natural languages are compositional in the respect that the meaning of a sentence is composed out of the meaning of its component words. Hence, the meaning of a sentence is exhaustively determined by the meaning of its component words and the syntactic structure of the sentence.[5]

Returning to classical computers, then, we can say that the symbols they manipulate belong to a language-like system. Hence, they are complex structures built out of basic symbols (the analogues of words) by means of the application of recursive syntactic rules. When a particular physical machine is a classical computer, the language-like symbols it manipulates will take a particular physical form. More specifically, there will be mapping from the symbols of the language onto types of internal physical states which the machine is capable of instantiating that specifies how the language is physically implemented or realized in the machine.[6] This mapping will pair each basic symbol of the language with a distinct type of internal state. With respect to complex symbols, the mapping will reflect that relating to basic symbols in the following way. Suppose a complex symbol C contains the basic symbols A and B that are mapped onto, respectively, internal state types S1 and S2. Then C will be mapped onto a complex state type that contains S1 and S2 as its proper parts. Moreover, syntactic relations between symbols will be mapped onto physical relations between the components of complex internal states. In other words, the mapping will be such that the symbolic structure of symbols is reflected in the physical structure of the physical states with which they are paired.

Different machines can manipulate symbols of one and the same language despite the fact that the way the symbols of the language are physically implemented or realized in one machine differs from how they are physically implemented or realized in another.[7] In other words, the symbols of a given language are multiply realizable at the physical level. When we consider a natural language such as English, this shouldn't be all that surprising. For any given word or sentence of English can be realized by a mark on a piece of paper when written down or by a sound when vocalized.

Classical computers manipulate symbols of the relevant language in generating output from input. This will typically involve such operations as copying symbols, comparing symbols, storing symbols, deleting symbols, retrieving symbols from memory, and so on. Consequently, intermediary symbols between the input and output are often generated in the course of computation. Manipulating symbols involves applying symbol-manipulating rules, and the particular rules that a machine applies in generating output from input constitute the program that the machine runs. Such rules will advert to the syntactic properties of symbols rather than any meaning they have, and so their application will not require any understanding of the symbols on the part of the machine. Consider an example. A machine takes sentences of English as input and produces sentences of English as output. One of the rules that it applies is this:

From any two symbols of the form 'if A then B' and 'A' produce the symbol 'B' as output.

Applying this rule to the input 'if today is Tuesday then tomorrow is Wednesday' and 'today is Tuesday' will generate the output 'tomorrow is Wednesday'. Applying it to the input 'if Pierre lives in Paris then he lives in the capital of France' and 'Pierre lives in Paris' will generate the output 'Pierre lives in the capital of France', and so on. In applying this rule, the machine doesn't need to understand the sentences it manipulates; rather, all that is required is that it is sensitive to their syntactic form.

Now of course, being sentences of English, the symbols the machine manipulates do have meaning, and the nature of the rule is such that, in applying it, the machine generates output which, in semantic terms, is coherently related to the input. More specifically, the output logically follows from the input in the respect that, if the input sentences are true, then the output sentence would have to be true. In other words, the machine is reasoning logically, for it is applying the logical rule of *modus ponens*. This example reveals the aptness of John Haugeland's (1981: 23) oft-quoted comment that the core assumption of classical computationalism is that, 'if you take care of the syntax, the semantics will take care of itself.'

The rule the machine applies constitutes one element of the program it runs. With respect to such rules we can draw a distinction between those that are hard-wired into the machine and those that are explicitly represented within it (Dennett, 1983). In the latter case, the rule will itself be encoded by means of a symbol akin to an input or an output symbol that is stored in the machine's memory, and that symbol will be causally implicated whenever the rule is applied. In effect, the rule will be an explicit instruction that tells the machine what to do with the input symbols it receives. Once again, though, following this instruction will be a mechanical affair that does not involve understanding the rule; rather, in responding to the rule the machine is merely sensitive to its syntactic properties. Familiar examples of such explicitly represented rules are those that constitute the word-processing programs run on contemporary PCs and Macs and some of the symbols on the tape of a Universal Turing Machine. But if an explicitly represented rule is going to drive a machine's symbol-manipulating behaviour, then the machine must 'know' how to apply the rule. Now in principle it could consult another explicitly stored symbol, which would, in effect, be a rule for applying a rule. But this can't go on forever; the machine's rule-applying behaviour must ultimately bottom out in behaviour that is not driven by an explicitly stored rule; at bottom, the machine's behaviour must be driven by rules that are hard-wired into it.

An important point to note in all of this is that there is a key distinction between two particular kinds of rules relating to classical computers. On the one hand, there are the syntactic rules of the language that the computer manipulates. These rules determine how the basic symbols of the language can be combined to create more complex symbols such as sentences. On the other hand, there are the rules that constitute the programs that the machine runs. These rules instruct the machine how to respond to input symbols. Thus, it is perfectly possible to have two separate machines that manipulate symbols of one and the same language while manipulating those symbols in quite different ways because they run distinct programs. For example, suppose that two machines both take pairs of Arabic numerals as input and produce Arabic numerals as output. The first machine always produces the numeral that represents the number that is the sum of the two numbers represented by the input numerals. The second machine, on the other hand, always produces the numeral that represents the number that is the product of the numbers represented by the input numerals. Thus, the first machine is an addition machine running an addition program and the second machine is a multiplication machine running a multiplication program.

This distinction between the syntactic rules of a language and the processing rules that belong to a program serves to undermine a line of argument developed by Susan Schneider (2011) in her defence and elaboration of classical computationalism. To see what this argument comes to, we need to understand the type–token distinction. A token of a type is a particular instance of that type. For example, the chair I am currently sat upon is a token of the type *chair*. The chair I usually sit on in my favourite café is another token of the type *chair*, as are all the other particular chairs in the world. The type–token distinction applies to symbols just as much as it does to furniture. For example, the printed word that began the previous sentence is a token of the word type *the*, and were you to read that sentence out aloud you would begin by producing another token of that same type.

Schneider argues that it is incumbent on the advocate of classical computationalism to provide an account of the nature of symbols, an account that explains why any two token symbols that are instances of the same symbol type are instances of the same type rather than two distinct types. She complains that such an account is not present in the literature and sets about remedying this potentially disastrous situation by providing her own account. According to Schneider's account, what binds together tokens of a given symbol type is that they have a common causal role within the machine, and this boils down to their being processed in just the same way. She argues that this account falls out of the core classical computationalist idea that computers are sensitive only to the formal or syntactic properties of the symbols that they manipulate. For, if two token symbols are syntactically distinct, then they must be processed differently by the machine, and if there is no such difference then, as far as the machine is concerned, the tokens can't be discriminated and so must belong to the same symbol type.

What I think is right about Schneider's line of thought is that, if a computer at a given point in time treats – or would treat – two token symbols in the same manner so that they have just the same causal role within the machine, then that is good grounds for regarding the symbols as tokens of the same symbol type. However, we need to be able to make sense of how two distinct computers (or two distinct time slices of the same computer) can manipulate symbols of one and the same language and yet do so quite differently because they are running non-equivalent programs. An example of this is provided by the addition and multiplication machines described above. This is a possibility that Schneider's account rules out because, if two computers run different programs with respect to symbols of a common language, those symbols will have a particular causal role in one of the computers and a quite different causal role in the other. In short, Schneider misses the distinction between the syntactic rules of a language and the processing rules of a program.

We are now in a position to see what the classical computationalist view of cognition comes to. Cognitive capacities such as perception, high-level reasoning, action planning, object recognition, language mastery, and so on, are underpinned by the activities of a classical computer or ensemble of classical computers that are housed in the brain. These computers manipulate representations that are language-like structures realized by brain states, and they do this by means of the application of rules. Applying these rules is a mechanical process that requires the neural-computer to be sensitive only to the syntactic properties of the representations that it manipulates. The upshot of all this is that the cognitive scientist's goal of explaining cognition involves identifying the language-like representational system employed by the computers housed in the brain and the rules that they apply to such representations whenever we exercise our cognitive capacities.

A framework for thinking about cognitive scientific explanation in the light of the core commitments of classical computationalism that has proved to be very influential has been developed by David Marr (1982). According to Marr, when there is a cognitive capacity that is a proper object of cognitive scientific study, then the exercise of that capacity will be underpinned by the operations of a computer housed in the brain. A complete account of the cognitive capacity will have three levels. The uppermost level is the computational theory. This specifies the semantic or intentional details of the computer's activity by indicating what information it takes as input, what it produces as output, and what intermediary information it produces in generating output from input. It also indicates why this information is relevant and identifies any mathematical operations the computer executes in its information-processing activities. The second level is the theory of representation and algorithm which specifies the syntactic details of the computer's activity. Thus, it identifies the particular language used by the computer and the symbol-manipulating rules it applies to the symbols of that language. The third level is the theory of hardware implementation. This provides the details as to how the higher-level symbol-manipulating activity is physically implemented or realized in the computer and, hence, the brain. For Marr, a given computational theory is consistent with many distinct theories of representation and algorithm, as the activity that the former characterizes is multiply realizable at the syntactic level. Similarly, a given theory of representation and algorithm is consistent with many distinct theories of hardware implementation, as syntactic activity is multiply realizable at the physical or neural level. Marr also implies that the cognitive scientist must begin at the uppermost level and work downwards, as one cannot uncover the details of, say, a computer's syntactic activity until one has a substantial understanding of its activity at the semantic, information-processing level.

I now come to an important distinction that needs to be engaged with. This is the distinction between the personal and the subpersonal, a distinction that has been highlighted by Daniel Dennett (1969, 2013). Personal-level states and processes are states and processes of a whole person rather than any of their components. For example, it is I, the whole person, who believe that Paris is beautiful in the spring and who executed the thought process of concluding that Pierre lives in Paris on the basis of being told that he lives in the capital of France and my long-standing belief that Paris is the capital of France. However, it would appear that there are some states and processes going on in my mind that are not had or done by me in this way. For example, suppose I hear the structurally ambiguous sentence 'the man chased the dog with the stick' and understand that it is the dog that had the stick rather than the man. What underlies this under-standing (a personal-level state) is a process of parsing the sentence, of determining its individual component words and identifying the syntactic relationship between those parts. This is something that, presumably, takes place in my brain but is done not by me but by a neurally realized compo-nent of me. In other words, the process of parsing a sentence takes place at the subpersonal level.

The personal–subpersonal distinction is not the same as the conscious–unconscious distinction. Perhaps all subpersonal states and processes reside beneath the level of consciousness, but there are plenty of uncon-scious personal-level phenomena; for example, any unconscious belief or desire would be an example of the latter.

The existence of this distinction between the personal and the sub-personal raises an important question for any advocate of classical computationalism: are they claiming that personal- or subpersonal-level cognitive processes are computational? This question has several dis-tinct possible answers: one could argue that subpersonal processes are computational while personal-level ones are not; or one could argue that personal-level processes are computational while subpersonal ones are not; or one could argue that they are both computational.[8] Historically, classical computationalists have rarely been clear on their answer to this question and tend not to be explicit with respect to which level they take themselves to be working at. Nevertheless, there are a couple of influen-tial and historically important computationalist theories where the level of cognition at which the theory is pitched is clear-cut, which I will now describe.

The first is David Marr's (1982) theory of vision, which attempts to char-acterize the subpersonal processes underlying our visual capacities. Marr portrays that process as consisting of three stages.[9] The input to the visual process is a two-dimensional retinal image that represents the wavelength and intensity of light falling on each of the many light-sensitive cells that make up the retina. The first stage involves generating a primal sketch from the retinal image. This is a representation of the location of sudden changes in light intensity across the retinal image, changes that tend to be caused by, and correspond to, significant features of the external viewed scene, such as object edges, boundaries between objects, changes in surface colour and texture, surface contours, and such like. The second stage involves generating a 2½-D sketch from the primal sketch. This represents the relative distance from the viewing subject of each point on the surfaces of viewed objects, together with their orientation relative to the subject. In the final stage a 3-D representation is generated from the 2½-D sketch which is an object-centred representation of the objects in the viewed scene that represents them in the manner of a stick figure.[10]

A classical computationalist approach firmly pitched at the personal level is Jerry Fodor's (1975, 1978, 1987) theory of propositional attitudes, such as beliefs and desires, and thought processes involving those states. According to Fodor, propositional attitudes are computational relations to representations that are realized in the brain. The representations involved here belong to a language, the language of thought (LOT for short). For Fodor, this is not a natural language such as English, Italian, Tagalog or Walpiri but, probably, one that is shared by all humans in being part of our distinctively human biological endowment. Hence, for example, believing that Paris is beautiful in the spring involves standing in the computational relation constitutive of belief to a token of the LOT analogue of the English sentence 'Paris is beautiful in the spring'. Desiring that Paris is beautiful in the spring involves standing in the computational relation constitutive of desire to a token of the LOT analogue of the English sentence 'Paris is beautiful in the spring', and so on for all the other distinct types of propositional attitude, such as intention, expectation, hope, fear, and such like.

The computational relation that one bears to a token of an LOT sentence comes down to the causal role of the sentence. Whether or not a token sentence of LOT expresses a belief, or a desire, or an intention, or whatever, depends upon how it is processed by the in-head computational mechanisms that have access to it and, thus, by its role in one's internal cognitive economy. Sentences that express beliefs are processed in the distinctive way that is characteristic of the belief relation, sentences that express desires are processed in the distinctive way that is characteristic of the desire relation, and so on. A popular and helpful metaphor due to Stephen Schiffer (1987) is that of belief boxes, desire boxes, and the like. Thus, when one believes that p, one has a token of an appropriate LOT sentence stored in one's belief box, and when one desires that p, one has a token of the same sentence stored in one's desire box. A sentence stored in the belief box will be processed by the computational mechanisms that access it differently from how a token of the same sentence stored in the desire box is processed. For example, suppose I desire a bottle of elderflower pressé and believe that there is a bottle of elderflower pressé in the fridge. Then the computational mechanisms that access the relevant LOT sentences in my belief and desire boxes will generate a token of the LOT analogue of 'I go to the fridge to get a bottle of elderflower pressé' that is placed in my intention box. This sentence in the intention box is subsequently processed in such a way as to cause me to go to the fridge. Now suppose that, although I wanted there to be a bottle of elderflower pressé in the fridge (and so had a token of the relevant LOT sentence in my desire box), in actual fact I didn't believe that there was. Then the computational mechanisms accessing the contents of my belief and desire boxes would not find anything in the former that, given what they find in the latter, causes them to generate an intention to go to the fridge to get a bottle of elderflower pressé. Thus, the LOT analogue of 'there is a bottle of elderflower pressé in the fridge' plays a different causal role in one's cognitive life when stored in one's belief box than it does when stored in one's desire box.

That completes my account of classical computationalism. I now turn to the task of evaluating that approach, as it has proved to be very controversial.

3 Evaluating classical computationalism

Jerry Fodor has done more than most to defend classical computationalism, especially with respect to his theory of propositional attitudes and the thought processes involving them. Fodor (1987) argues that classical computationalism explains several key features of our cognitive lives that would otherwise be mysterious. The first feature is that thought is productive in the respect that there are limitless distinct propositional attitudes that we humans are capable of instantiating. Moreover, we routinely instantiate propositional attitudes that we have never instantiated before. For example, if you read the sentence 'aardvarks aren't very good at peeling mangoes with a potato peeler', you will readily come to entertain a thought – namely, the thought that aardvarks aren't very good at peeling mangoes with a potato peeler – that you have never entertained before. The language-like structure of LOT explains the productivity of thought, argues Fodor, as, on account of the recursive nature of its syntactic rules, a language with only finitely many basic symbols and finitely many rules will have infinitely many distinct sentences.

The second feature of our cognitive lives that needs explanation is the systematicity of thought, the fact that anyone capable of thinking that object x stands in relation R to object y (whatever x, y and R might be) is also capable of thinking that y stands in R to x (and vice versa). For example, anyone capable of thinking that the dog chased the postman is also capable of thinking that the postman chased the dog (and vice versa). Once again, the language-like structure of LOT explains this as the very resources – in terms of basic symbols and rules – that are employed in thinking that x stands in relation R to y are employed in thinking that y stands in relation R to x.

The third feature of cognition that requires explanation is its rational coherence. Typically, when a person reflects upon a collection of her propositional attitudes, she will come to token further propositional attitudes that are coherently related to that collection in terms of their content. For example, if I reflect on my desire for a bottle of elderflower pressé, and my belief that there is a bottle of elderflower pressé in the fridge, I will typically come to form the intention to go to the fridge for a bottle of elderflower pressé. As classical computers are devices that are programmed in such a way that their output symbols are coherently related to their input symbols in terms of their meaning or content, portraying cognition as a form of classical computation readily explains its rational coherence. This argument also applies to subpersonal cognitive processes such as those involved in perception, language use, object recognition, and such like. For, if such processes are to support our cognitive capacities, they must be rationally coherent. For example, we would hardly be able to see if subpersonal processes did not generate output representations of the shape, size, colour, location and movement of objects in the viewed scene from input representations concerning how those objects stimulate the retina.

Nevertheless, classical computationalism has come under considerable attack in recent years, and I will now discuss some of the most interesting objections.[11] Paul Churchland (2007, 2012) objects that classical computationalism involves a fixation on the thoughts we express by means of language and an illegitimate tendency to view humans as being fundamentally different from other animals.[12] Churchland begins by noticing that many non-human animals cognize: they perceive the world, acquire knowledge about it, and act on that knowledge in such a way as to satisfy their needs. However, it would be absurd to attribute to such animals discrete propositional attitudes of the kind that are well suited to explanation by appeal to a language of thought. Churchland adopts this view of non-human animals as a result of 'the extraordinary difficulty encountered in trying to teach any of them the normal use of human *language*, the original and still prototypical system for the expression of propositional attitudes' (2007: 90). As it stands, this might be taken to suggest that human and non-human cognition are fundamentally different and, reflecting this, that classical computationalism should be restricted to theorizing about the former. However, Churchland continues by arguing that much of the cognitive activity of humans is akin to that of non-human animals in that it is not bound up with language or typically amenable to linguistic expression, and he provides the examples of operating a lathe and driving. Moreover, to place a fundamental divide between humans and their non-human relatives is to ignore the evolutionary fact of continuity between the different species.

I'm not convinced by this argument for the following reasons. First, the inability of non-human animals to learn language, something that humans do with ease, requires explanation. The classical computationalist has a ready explanation: the mind-brain of humans, unlike those of their simpler cousins, deploys a language-like system of representation that is perfectly designed for learning and using a language. Hence, language mirrors the human mind (Pinker, 2007; Boeckx, 2010). Therefore, in portraying humans as being different from non-human animals with respect to language, Churchland's argument may have backfired. For perhaps the reason why the latter cannot acquire language is because they are fundamentally different from us. Second, Churchland wants to put the thoughts that we express linguistically at the periphery of human cognition, but it is far from clear that this is a legitimate move. We humans are surrounded by language throughout our waking lives and are constantly expressing our thoughts to our fellows by means of language and acquiring new thoughts as a result of encountering sentences produced by others. Without such linguistic interchanges our knowledge base and cognitive skills would be significantly diminished.

Third, with respect to the cognitive activities that Churchland claims do not involve discrete propositional attitudes and are not typically linguistically expressed, matters are not as clear-cut as he implies. When I'm driving I am rarely called upon to express what I'm doing linguistically, but perhaps I could if I wanted to: 'I'm slowing down now so as to get round the upcoming sharp bend without crossing over onto the opposite side of the road.' If I did find it difficult to keep up this linguistic expression for any amount of time, that might not have anything to do with the form of the cognitive states and processing involved in driving. Perhaps the contents of those states are so fine-grained and specific that I haven't learned the appropriate words to enable me to express accurately the structured representations I token. After all, it is no part of classical computationalism that we have a word in the public language that we speak corresponding to every representation we are capable of instantiating in cognition. Alternatively, driving is very demanding in terms of attentional resources, so perhaps it places too much of a processing demand on us to express linguistically our cognitive states while driving.

Fourth, Churchland needs more by way of argument to support his contention that the cognitive lives of non-human animals do not feature the manipulation of representations that have a language-like structure. For example, C. R. Gallistel (1999) has developed a classical computationalist explanation for the ability of insects to navigate by means of dead reckoning.[13]

A second line of objection to computationalism is suggested by research into the reasoning capacities of humans. We have seen that the rationality of humans and the logical coherence of our thought processes have been appealed to in order to motivate classical computationalism. Indeed, some of the most prominent early exercises in Artificial Intelligence involved attempting to program computers to reason logically in a manner that reflects the way that humans reason.[14] But, it might be argued, such considerations would be undercut if it turned out that humans were far from rational and that much of our 'thinking' relied on the use of heuristics or rules of thumb that often lead us astray rather than on logical rules such as *modus ponens*. Indeed, over the past few decades a rich body of research has uncovered the stark limitations of our reasoning powers, particularly with respect to reasoning with conditionals ('if–then' statements) and reasoning to do with probability and statistics. For example, consider the following cases.

In his book *Thinking, Fast and Slow*, Daniel Kahneman (2012) recounts the findings of his work with his long-time colleague Amos Tversky, who used experimental techniques to uncover failings in human reasoning. One classic example relates to judging how probable it is that a particular proposition is true. Suppose that one has two propositions, P and the more complex P and Q. It is a basic law of logic that the probability that P and Q is true is not greater than the probability that P is true. This is an upshot of the fact that P and Q cannot be true without P being true. Nevertheless, claims Kahneman, people often ignore this fact in their reasoning. In one experiment subjects were asked to judge the probability of various propositions about a woman called Linda who was described to them in the following terms:

> Linda is thirty-one years old, single, outspoken, and very bright. She majored in philosophy. As a student, she was deeply concerned with issues of discrimination and social justice, and also participated in anti-nuclear demonstrations. (2012: 156)

The list of propositions presented to the subjects of the experiment included these:

1   Linda is a bank teller.
2   Linda is a bank teller and is active in the feminist movement.

Despite the fact that the above-mentioned law of logic implies that 2 cannot be more probable than 1, 85 per cent of the subjects in the experiment judged 2 to be more probable than 1.

Another kind of case Kahneman documents is that in which an individual's judgement of the probability of a proposition is driven by stereotypical knowledge and completely ignores information relating to frequency. For example, subjects are told that a man called Steve has been drawn at random from a representative sample of the population. They are then told that he is shy and withdrawn, values order, and has little interest in people or the real world. Finally they are asked whether Steve is more likely to be a librarian or a farmer. Most subjects answer that Steve is more likely to be a librarian, thereby ignoring the fact that, given that there are twenty times as many farmers as there are librarians in the USA (where the experiment was carried out), Steve is more likely to be a farmer even if a higher proportion of librarians than farmers fit his profile.

In these and many other cases, argues Kahneman, we fail to reason logically. Instead, we rely on heuristics or rules of thumb and our reasoning is subject to biases. In the case of Steve, the heuristic relates to resemblance; we judge him more likely to be a librarian because that proposition better resembles our knowledge of the stereotypical librarian than it does the stereotypical farmer.

Kahneman also argues that, in cases where we have a decision to make between two options, which one we choose can be influenced by the way the options are 'framed'. For example, in an experiment subjects are given $50 and the choice of gambling to keep the money, at the risk of losing it all, or settling for a 'sure outcome', which involves handing back $30 and keeping $20. The sure outcome can be described or framed in different ways, as involving keeping $20 or losing $30. What Kahneman discovered is that people tend to prefer the sure outcome to the gambling option when the former was described as keeping $20 than when it was described as losing $30, despite the fact that the situations are logically equivalent. In other words, when deciding between possible options, humans are biased against options that they see in terms of losing.

Gerd Gigerenzer (2014) is another important psychologist who emphasizes the role of heuristics in our cognitive lives. However, he doesn't endorse as negative a view of heuristics as Kahneman. Rather, he thinks that the employment of heuristics is not generally inferior to logical and statistical reasoning, and in conditions of uncertainty they can be very useful in driving adaptive behaviour.

Fascinating and important though these studies are, I'm not convinced either that they show us to be irrational or that they thereby undermine classical computationalism. They would be more of a challenge if it was shown that humans were incapable of reasoning logically, but the fact that we sometimes fail to reason logically doesn't imply that. There is a family of approaches to reasoning, known as dual aspect theories, which postulate two distinct systems or processes that are involved in reasoning.[15] One of these is involved when we are called on to make decisions quickly in less than ideal circumstances. Here we rely upon heuristics, and our reasoning often fails to obey the laws of logic. The second process demands much more of us in terms of mental effort, attention, time and information, and when we are able to execute it we come much closer to reasoning in a logical manner. In fact Kahneman himself is an advocate of a dual-reasoning approach, distinguishing between system 1 and system 2. System 1, the home of heuristics and biases, 'operates automatically and quickly, with little or no effort and no sense of voluntary control' (2012: 20). System 2, on the other hand, allows us to be more logical and 'allocates attention to the effortful mental activities that demand it, including complex computations' (ibid.: 21).

This distinction between two processes or systems associated with reasoning suggests a defence of classical computationalism to the effect that Kahneman's system 2 and its kin fit well with that theory of cognition. But, if we left it at that, the implication might be that classical computationalism could explain only a limited aspect of our cognitive lives. But perhaps such a retreat is not mandatory. Reasoning involving heuristics, though it might fail to honour the laws of logic, is hardly random and undisciplined, and if we do apply heuristics routinely then what is needed is some explanation of how and why we do so. Classical computationalism may well have the resources to provide such an explanation, as it does seem possible to program a computer to apply heuristics. For example, when presented with problems like that in the case of Steve, one could program a computer to draw solely on stereotypical information when statistical data was not at hand. With respect to the framing bias, that case involves representing the options as involving either losing or keeping. Now one could easily program a computer with a rule such as 'avoid losing at all costs' which kicked in whenever an option was represented as involving loss.

With respect to the case of Linda the bank teller, it is not obvious that the subjects are thinking illogically. Suppose the subjects have a knowledge of what a typical bank teller is like (namely that they are apolitical or politically conservative) and believe that people tend not to change in their deep-seated political opinions once they have reached college age. Then one possibility is that they understand the statement that Linda is a bank teller as implying that Linda is a stereotypical bank teller, so that their choice is between deciding whether it is more likely that Linda is a stereotypical bank teller or one who is an active feminist.[16] Given what they have been told about Linda and their belief that people tend to retain their political commitments, the more rational thing to do would be to judge the statement that Linda is a bank teller and is active in the feminist movement is more likely to be true.

A third objection to classical computationalism relates to the meaning or content of representations. Consider a standard human-manufactured classical computer such as the laptop I used to write this book. I will call such machines artefactual computers to distinguish them from natural computers such as the brain. Its inputs, outputs and internal states do have meaning or content, but that meaning comes from us; it is a product of our interpretation of the computer. In other words, the representations in which artefactual computers traffic have what is sometimes called derived as opposed to original intentionality[17] (Searle, 1980; Haugeland, 1981). Our mental states, on the other hand, have original intentionality.[18] As interpreting a symbol involves adopting a relevant mental state towards that symbol, to hold that the meanings of our mental states are based upon interpretation threatens an infinite regress. For lying behind each interpreting mental state would have to be a further mental state of interpretation to fix its meaning. So, the argument concludes, there is a fundamental difference between artefactual computers and us or our mind-brains.

In response to this argument, one might object that, just because the meaning of the symbols manipulated by artefactual computers is derived, it doesn't follow that the same applies to all computers. Thus, in 'natural' computers such as the human mind-brain, the representations manipulated have original intentionality. However, this line of thought generates a challenge: namely, that of explaining this meaning without any appeal to acts of interpretation. The project of constructing such an explanation is generally known as that of naturalizing content. There are various naturalistic theories of content on the market, and most of them fall into one of three broad types. First, there are informational theories, according to which the meaning of a mental representation is a matter of what typically causes its tokening (Fodor, 1987, 1990; Dretske, 1981, 1988). For example, a representation would mean *dog* where it is the case that it was caused to appear in response to the presence of a dog but never in response to a non-dog. Second, there are teleological theories that ground the meaning of mental representations in their biological function, which is in turn a matter of their evolutionary history (Millikan, 1984, 1993; Papineau, 1993). For example, a representation would mean *dog* if it were its biological function to occur in the presence of a dog. Third, there are functional role theories, according to which the meaning of a mental representation is a matter of the relations that it bears to other representations (Block, 1986; Harman, 1987). For example, a representation would mean *dog* if it belonged to a network of representations including ones that meant *bark, animal, quadruped,* and so on, and bore the appropriate relations to those representations. All of these theories have their problems,[19] but I see no reason for concluding that a viable naturalistic theory of content will not be forthcoming.[20]

4 Connectionism

I now come to the second version of computationalism. This is known as connectionism or Parallel Distributed Processing; it came to prominence in the 1980s and constitutes a major contemporary alternative to classical computationalism.

Although classical computationalists conceive of cognition as being realized in the brain, their approach wasn't historically motivated by study of the brain. Moreover, describing cognition in terms of the rule-governed manipulation of language-like symbols is not to describe it in such a way as to gel directly with the way neuroscientists describe brain activity when they talk of the electro-chemical interaction between millions of neurons, each of which is connected to thousands of other neurons. The same is not true of connectionism, as it portrays cognition as being based upon the activity of networks of simple units that bear considerable similarities to the networks of neurons we find in the brain. Hence, connectionism is often described as a neurally inspired approach (P. M. Churchland, 2013).

Connectionist networks[21] are networks of simple units that are connected to one another. The activity of such a unit (often called a node) is influenced only by local factors. Such activity is the bedrock of processing within the network, and at any point in time such local processing can be taking place at numerous different locations in the network. Hence, connectionist networks are described as parallel as opposed to serial processors. Connectionist networks come in a variety of forms,[22] but the most widely discussed is the three-layered feed-forward system. Represented diagrammatically such a network looks like figure 2.2.

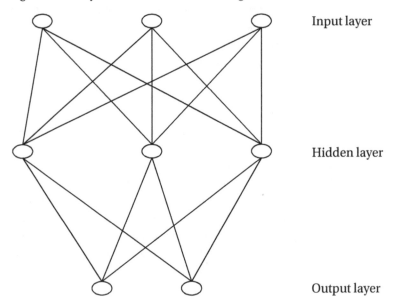

Input layer

Hidden layer

Output layer

Figure 2.2

The network units are arranged into three layers, namely, an input layer, an intermediate (or hidden) layer, and an output layer. Each layer can have any number of units – that is, it doesn't have to have the same number as in figure 2.2. At any point in time each unit will be in a state of activation, a state that has a numerical value. In some cases the unit can be either active (1) or inactive (0), but in more complex cases there are a range of levels of activation between being fully on and fully off, each of which is represented by a number between 0 and 1. Each unit requires a certain amount of stimulation in order to become active; this is its threshold value. Each unit in the input layer is linked by a number of connections to many units in the intermediate layer, and the intermediate units are similarly linked to units in the output layer. Impulses are transmitted along these connections. The strength of an impulse emitted by a unit – which has a numerical value and constitutes its output – is a function of its level of activation. The strength, and hence the numerical value, of the impulse received by the receiving unit is a product of the strength of the initial impulse and the weight of the connection. For example, if a unit sends an impulse of value 1 down a connection of weight 2, the impulse received will be of value 2.

The connections between units fall into two classes, namely, excitatory and inhibitory. Input received via an excitatory connection pushes the receiving unit towards a state of activity or excitement, whereas that received via an inhibitory connection has the opposite effect. The level of stimulation a unit receives as input is the sum of the values of the impulses it receives via excitatory connections minus the sum of the values of the impulses it receives via inhibitory connections. When units in the input layer are stimulated, impulses pass along connections to the intermediate layer, so stimulating activity there. Impulses are then passed to the output layer, resulting in patterns of activity at that level. Consequently, the system transforms patterns of activation at the input layer into patterns of activation at the output layer. As these patterns of activation can be represented by a list of numbers, with each number specifying the activity level of a particular unit, the network transforms vectors into vectors.

The specific input–output behaviour of the network is determined by the nature and weight of the connections and the threshold values of the units. Adjusting the connection weights will alter the network's input–output behaviour. To make this a little more concrete, consider the behaviour of a portion of the above network, namely the portion linking the input nodes to the central node in the hidden layer. Figure 2.3 represents this diagrammatically.

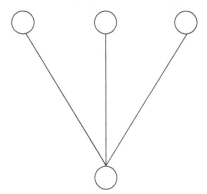

Figure 2.3

Suppose that the relevant details concerning this portion of the network are as follows. With respect to the connections, the leftmost is excitatory and of weight 2, the middle connection is inhibitory and of weight 1, and the rightmost is excitatory and of weight 3. Further, the hidden layer node has a threshold value of 5, and all the units can either be on/active (1) or off/inactive (0). When an input layer node is on/active, it sends an impulse of value 1 down all connections leading from it. When it is off/inactive, it sends no impulse at all (that is, an impulse of value 0). Now suppose that initially the hidden layer unit is off/inactive. Then the activity pattern <1, 0, 1> at the input layer will result in the hidden layer node becoming active (1). This is because the input delivered via the outer excitatory connections will come to 5 (as $(2 \times 1) + (3 \times 1) = 5$) and that delivered by the middle inhibitory connections will be 0, giving a net input of 5, a figure meeting the threshold value. Any other pattern of activity will not have this effect – that is, the hidden layer unit will remain off/inactive. So this portion of the network maps <1, 0, 1> onto <1> and all other vectors onto <0>. This mapping could be altered by changing the connection weights. For example, if the leftmost connection weight was increased to 3, then the input pattern <1, 1, 1> would be mapped onto <1> by the network, as the net input delivered by the three connections would be $(3 \times 1) + (3 \times 1) - (1 \times 1)$, which, of course, equals 5.

For the connectionist the mind-brain is a connectionist network or ensemble of such networks. But one might ask how this could be so. How could a system like the one described above support cognition? The crucial point is that patterns of activation at the input and output layers can have semantic properties – that is, they can represent. Another important feature of such networks is that they can learn by changing the profile of their input–output behaviour in the light of experience. To explain this, it will be helpful to discuss a particular connectionist network that has been examined in great detail by Paul Churchland (2012), one of the most significant philosophical champions of connectionism.

The network in question is Cottrell's (1990) face-recognition network. The input layer consists of 4,096 units arranged into a 64 × 64 unit grid. The input layer is stimulated by black and white images of faces, with each unit being sensitive to the lightness/darkness of the image at a specific point. Hence, each potential distinct input pattern represents a particular distinct possible image. The input units are connected to a series of eighty units at the middle layer, which are in turn connected to a series of eight units at the output layer. Now consider that output layer. The designer of the network attributes meaning to particular patterns of activation at that level. The leftmost unit is interpreted as specifying whether the image stimulating the input layer is that of a face or not. Thus, if it responds by becoming highly active, it is delivering the judgement that the image is of a face. However, if that unit is inactive or active to only a minor degree, then that output is delivering the judgement that the image is not of a face. A mid-range response could be taken as constituting a non-committal 'don't know' response. Moving rightwards, the next two units of the output layer relate to the sex of any face image presented to the input layer. A highly active response of the second unit constitutes a judgement that the image is of a male and a low response constitutes a judgement that the image is not of a male. The next unit, the third, relates to femalehood, with a highly active response representing the face image as of a female and a relatively inactive response representing it as not of a female. The remaining five units serve to represent the names of particular individuals. Thus, for example, one particular pattern of activity over those units will encode the name 'Mary', so that when it is instantiated it will represent the face image as being of Mary. Another will encode the name 'Frank', and so on for a range of individuals. In sum, then, what the network does is transform patterns of activation that serve to describe black and white images into patterns of activation, specifying whether the image in question is of a face and, if so, if it is the face of a male or a female and, moreover, what is the name of the bearer of the face.

Like many connectionist networks, this face-recognition network has the ability to learn. Initially, the connection weights are set randomly. As a result, the network makes many mistakes in, for example, misidentifying the sex and identity of face images presented to it. But the network is subjected to a training process during which it slowly improves its performance by having its connection weights incrementally altered. Here is how such a process goes. Suppose that the network is stimulated by an image of Mary and its response is to deliver an output that specifies the image as being of a male and of Frank. Then that output is compared with the correct output. Suppose the correct output was <1, 1, 0, 1, 1, 0, 0, 0>, whereas the actual output was <1, 0, 0, 1, 0, 1, 0>. As a result of comparing these two vectors, the connections leading to the output units are altered in the following manner. If a unit became active when it should have been inactive, then the inhibitory connections leading to that unit are slightly strengthened and the excitatory connections slightly weakened. Conversely, if a unit remained inactive when it should have become active, then the inhibitory connections leading to that unit are slightly weakened and the excitatory connections slightly strengthened. If, on the other hand, a unit performed correctly, then the connections leading to it are left unchanged. Making these adjustments will improve the network's performance with respect to that particular input. Then a second image is presented to the input layer and the connection adjustment process is carried out again in response to any error. A potential problem here is that the second run of adjustments might undermine that good work done on the first run of adjustments with respect to the first image. However, if a large number of images are presented to the network and the process of adjusting connections is continued for each one, then gradually the network will develop connection weights such that it will deliver correct answers for all images presented during the training period. Moreover, it will be able to deliver correct outputs for images that it did not encounter in the training period – for example, identifying an image of Frank from an unfamiliar angle as being of a male face that belongs to Frank. This process of gradual learning is known as back propagation.

This face-recognition unit traffics in representations, as the transitory states at the input and output layer – and, as we shall see, those at the middle layer – are meaningful. We might also say that the trained network stores knowledge. For example, it knows what Mary, Frank, and so on, look like and how a male face differs in appearance from a female face. But, one might ask, where is such knowledge stored? The natural answer is that all that knowledge is stored collectively by the connection weights. Hence, particular items of knowledge are not stored by means of distinct states that can be readily isolated from other items of knowledge. In this respect the network differs from a typical classical computer that might contain a distinct representation of Frank's appearance, on which it draws in identifying an image of Frank as such, and which is stored in a different location from the representation concerning Mary's appearance. A common way of characterizing this situation is to say that, in a connectionist network, knowledge is distributed throughout the network.

There is also a respect in which the transient representations of the face-recognition network at the input and output layers are distributed. Unlike sentences of a language such as English or the language of thought as conceived by Fodor, they do not have any internal syntactic structure. Moreover, distinct patterns of activation can involve the same units and can overlap to a certain extent. For example, suppose that the pattern of activation over the rightmost five output layer units that represents Frank is <1, 1, 0, 0, 0>, whereas that that represents Mary is <1, 1, 1, 1, 0>. Then those two distinct representations overlap with respect to the first, second and fifth unit.

Churchland (2012) places great emphasis on what happens at hidden layers of connectionist networks and argues that patterns of activation here have crucial representational significance and, indeed, serve to encode concepts.[23] His views here are important and merit extensive consideration. Churchland argues that distinct possible patterns of activation at a particular layer of a network stand in distance relations to one another, with some patterns being closer to one another than others. It is helpful to follow Churchland in explaining this idea geometrically in connection with networks much simpler than the face-recognition module we have been considering. Consider a network that has a middle layer made up of three units. Then any particular pattern of activation can be represented in terms of its position in a cube where the x axis relates to the activity level of the first unit, the y axis to the activity level of the second unit, and the z axis to the activity level of the third unit. For example, the pattern <0, 1, 1> would be located in the front, top, left-hand corner of the cube; <1, 1, 0> would be located in the front, bottom, right-hand corner; <0.5, 0.5, 0.5> would be located in the centre of the cube; and so on. When one conceives of activation patterns in this way, then it is easy to see them as having distance relations to one another where those relations are akin to distance relations between points in a physical cube. For example, <0, 0, 0> (located at the rear, bottom, left-hand corner of the cube) is closer to <0.5, 0.5, 0.5> (the centre point) than it is to <1, 1, 1> (front, top, right-hand corner). In short, then, activation patterns in a given layer are points in a multidimensional space where each point bears a distance relation to all other points. In the case of a layer with three units, the multidimensional space has three dimensions and so can be represented visually by means of a cube. In the case of a layer such as the middle layer of Cottrell's face-recognition network,[24] the space has many more dimensions and so cannot be represented visually. Nevertheless, all its possible patterns of activation are points in a multidimensional space, with each point bearing distance relations to all other points.

Now let's return to Cottrell's face-recognition network. As its input–output profile changes through learning, changes take place with respect to activity at the middle layer whenever the network is presented with a particular face image. Initially, images of a particular type of stimulus will provoke a 'noisy' response at the middle layer. For example, images of different male faces will provoke a range of responses that are scattered across the multidimensional space and will no more cluster together than the responses to a randomly selected collection of male and female faces. But, as the network learns, this changes, so that middle-level responses to distinct male faces come to cluster around a particular point in the space. Similarly, the middle-level responses to female faces come to cluster around another point in the space. These clusters constitute distinct non-overlapping subvolumes in the space where the distances between any two points in a given subvolume are shorter than the distances between any two points belonging to distinct subvolumes. The same holds, *mutatis mutandis*, for other categories of image – for example, images of a particular individual (such as Mary) or images of members of a particular family. Note that these subvolumes have a hierarchical organization so that subvolumes at one level can be contained within larger subvolumes at a higher level. For example, each of the subvolumes relating to individual females will be located within the 'female' subvolume, while each of the subvolumes relating to individual males will be located within the 'male' subvolume.

For Churchland, networks such as Cottrell's face-recognition network constitute good models of cognition; cognitive capacities such as face recognition are grounded in the workings of a connectionist network.

Now consider a network similar to the face recognition network that has been trained to recognize animals and so can serve as a potential model of a broader human object-recognition capacity. When stimulated by a black and white image of a particular type of animal, it tokens a pattern of activity at the middle layer that clusters around a particular point in multidimensional space, thus giving rise to a number of subvolumes each corresponding to a particular type of animal. These subvolumes will bear distance relations from one another. For example, the dog subvolume will be quite close to the cat subvolume, much closer than it is to the rat subvolume. However, the dog subvolume will be closer to the rat subvolume than it is to the lizard subvolume. The dog, cat and rat subvolumes will all be located inside a larger subvolume, namely the mammal subvolume, which does not contain the lizard, snake and turtle subvolumes. In contrast, these will all be located in the reptile subvolume.

According to Churchland, these subvolumes constitute concepts. When the network tokens an activation pattern in the dog subvolume, a subvolume of activation patterns typically caused by images of dogs, then it has conceptualized the image as being of a dog. What is important to notice is that different activation patterns will be instantiated on different occasions when the DOG concept is applied. Thus, in contrast to Fodor's language of thought approach – which conceives of concepts as being symbols in the language of thought so that, whenever a given concept is applied, one and the same mental representation is instantiated – application of one and the same concept can involve the instantiation of different representations (in the form of activation patterns) on different occasions.

Churchland invokes the term 'prototype' in connection with this view of concepts. A prototypical member of a category (or instance of the corresponding concept) is a central member of that category (or central instance of that concept). For example, a Labrador (but not a Pekingese) would be a prototypical dog and a trout (but not a turbot) a prototypical fish. The activation patterns at the centre of the subvolumes corresponding to concepts in the kind of networks that Churchland examines are typically instantiated in response to stimulation by prototypical instances of the concept in question. So, for example, in the animal-recognition network I have described, a Labrador (or, more accurately, an image of a Labrador) would cause a pattern of activation lying at the centre of the dog subvolume, whereas a Pekingese (or an image of a Pekingese) would cause a pattern of activation towards the edge of the subvolume.

As we have seen, connectionists such as Churchland are committed to the reality of mental representations. These take the form of patterns of activation that are meaningful. This raises the question of why these patterns of activation mean what they mean: what is the basis of their content? Of course the same question arises for the classical computationalist, and I have already described a range of potential answers. Many of those answers are available to the connectionist. Indeed, my description of the face- and animal-recognition networks might suggest the endorsement of a version of informational semantics by Churchland. According to that view, for example, activation patterns at the middle layer of Cottrell's face-recognition network that represent a face image as being of Mary have that content because they are reliably caused by images of Mary and not by images of any other individual. And activation patterns in the dog subvolume constitute the animal-recognition network's DOG concept because they are reliably caused by dogs, with their centre point being reliably caused by prototypical dogs. However, Churchland (1989, 2007, 2012) is insistent that he wants to reject informational semantics and endorse a holistic alternative that he calls 'state-space semantics' or 'domain portrayal semantics'.

In arguing for state-space semantics, Churchland compares concepts to maps. To see what this comes to, recall the point that some subvolumes in the animal-recognition network are closer to one another than others. For example, the distance between the cat and dog subvolumes is smaller than that between the dog and lizard subvolumes. As, for Churchland, our animal-recognition capacities are grounded in the workings of such a network in the brain, such distance relationships explain our judgements of similarity. For example, they explain why I judge dogs as being more similar to cats than they are to lizards and lizards as being more similar to snakes than they are to dogs. For Churchland, such distance relations are not a mere contingent feature of our animal concepts; rather, they are constitutive of those concepts in the respect that they are central to those concepts being the concepts that they are. Thus, one couldn't have the concepts DOG, CAT, LIZARD and SNAKE without those concepts having the distance relations described above. If an individual putatively regards dogs as being more similar to lizards than they are to cats and snakes as being more similar to dogs than they are to lizards, then what that implies is that they don't actually have the concepts DOG, CAT, SNAKE and LIZARD. For Churchland, then, by their very nature, concepts belong to networks of concepts the elements of which bear specific distance relations to one another. Hence, a pattern of activation encodes a particular concept (such as DOG) and so has the relevant meaning or content (such as *dog*) because of its distance relations to a whole range of other concepts. This is what is known as a holistic position because it portrays the content or meaning of a concept or mental representation as being determined by the entirety of its relations to a large collection of other concepts or mental representations.

Churchland does not only wish to endorse holism in arguing that content is a matter of distance relations between patterns of activation in a connectionist network. In addition, he seems to want to deny that causal connections between patterns of activation and the world outside of a network have played any role in determining content or meaning. This comes out in his discussion of maps, where he provides an account of their meaning and draws parallels between maps and connectionist networks. Consider a map of a particular terrain – say, the centre of Oxford. What makes that map a map of central Oxford? Churchland's (2012) answer is as follows. The map has certain features, such as lines, dots, blobs and other markings, that have distinct spatial relations to one another. To be a map of Oxford, these features must, in terms of their distance relations to one another, correspond to the distance relations between various objective features of central Oxford, such as buildings, roads, rivers, and such like. This correspondence need not be perfect, as some maps are more accurate than others, but if there exists a rough approximation then the map is a map of central Oxford. Similarly, argues Churchland, a network represents a particular objective domain (for example, the domain of animals) if the distance relations between its patterns of activation roughly correspond to the distance relations between the members of the objective domain. This perspective is captured in the following two representative quotations.

> [T]he meaning or semantic content of one's personal cognitive categories, whether innate or otherwise, derives not from any feature-indicating nomic relations that they may bear to the external world. Rather, it derives from their determinate place in a high-dimensional neuronal activation space, a space of intricate and idiosyncratic similarity relations, a space that embodies a highly informed 'map' of some external domain of lasting properties. . . . [T]he correct account of first-level learning requires us to put aside any form of atomistic, externalist, *indicator*-semantics in favour of a decidedly holistic, internalist, *domain*-portrayal semantics. (2012: 28–9)

> [The] source [of semantic content] is the relative position of the concept in a high-dimensional similarity space that also contains a large number of *other* concepts, which *family* of concepts collectively portrays (or purports to portray) the global similarity-and-dissimilarity structure of some external domain of features, kinds, or properties. Which and whether any of those concepts happens to be regularly *activated*, via the perceptual transduction of some external stimuli, is entirely incidental and inessential to the meaning or semantic content of that concept. (Ibid. 2: 93)

My objection to this position is that it is based on a mistaken view of the meaning of maps. Suppose that a student new to Oxford draws a map on the basis of her wanderings around central Oxford and subsequently uses it to help her successfully navigate that area. Then the map is a map of Oxford. Now suppose that someone else makes a physically indistinguishable map on the basis of exploring some other city distant from Oxford and subsequently uses that map to successfully navigate that city. Then it seems to me that the second map is not a map of Oxford. What this suggests is that, in order for something to be a map of a particular area, it is not sufficient that there exists a rough correspondence between its features and their distance relations, on the one hand, and those of the area in question, on the other. What is needed, in addition, is some kind of causal connection between the map and the terrain. This causal connection might involve the features of the map having been caused by interaction with the terrain. Alternatively, it could involve the map's causally driving an individual's navigations of the terrain.

If we apply this moral to the case of a connectionist network, the upshot is that a rough correspondence between the network and a particular domain is not sufficient for the network to represent that domain. In addition there must be some significant causal connection between the domain and the network. This could involve the network's mature state having been caused by interaction with the domain or the network's causally driving an individual's interactions with the domain. The issues here are complex and controversial, and I won't pursue them any further. However, I should stress a point that I've already implied: the viability of connectionism doesn't depend upon the strength of Churchland's semantic theory. As in the case of classical computationalism, there are a number of alternative theories that seek to explain content or meaning, and the failure of one does not imply the failure of all others. Indeed, some theories endorsed by classical computationalists can be adopted by the connectionist.

Thus far I have been focusing on three-layered feed-forward networks where stimulation flows in one direction only, from the input layer to the hidden layer and then to the output layer. However, a slightly different kind of network has received a lot of attention in recent years and, indeed, has been enthusiastically received by Paul Churchland. This is the simple recurrent network.[25] Such systems have a layer of units called the context layer in addition to the three layers we have already encountered. The pattern of activation here is a direct copy of the pattern of activation at the hidden layer during the previous cycle of processing and, along with the new pattern of activation at the input layer, constitutes the input to the hidden layer. In effect, the context layer constitutes a short-term memory that remembers the most recent pattern of activation at the hidden layer, so enabling the network to take into account its recent experience when processing new input. Figure 2.4 is a diagram of a simple recurrent network.

Churchland (2012) argues that the advantage of simple recurrent networks over standard feed-forward networks relates to their dynamism. They do not just process static input presented at a particular point in time. Rather, they take into account data relating to an extended stretch of

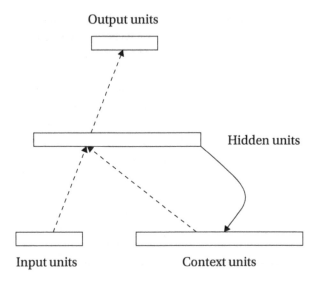

Figure 2.4

time. This means that, when the network is stimulated at time t, the output it generates will not be exhaustively determined by that input. For it will also be influenced by how the network responded to input presented at an earlier point in time. This, argues Churchland, makes simple recurrent networks particularly good at modelling perceptual processes, as perception is typically a dynamic process.

An interesting example here relates to the perception of colour. Suppose one has a network that is stimulated by light reflected off distal surfaces and whose units are sensitive to the wavelength of that light. The network has the job of determining the colour of the surfaces from which the light is reflected. The problem here is that a surface that has a particular colour will not always reflect light of the same wavelength. Rather, the wavelength of the light that it reflects will be different in different lighting conditions and so will change as lighting conditions change. This fact is well known to the human visual system and is reflected in the phenomenon of colour constancy, whereby a given surface will appear to have a constant, fixed colour as we view it across a range of different lighting conditions. For example, suppose on a sunny day you meet outside a café a friend who is wearing a bright red shirt. His shirt looks red to you. You then retreat into the dimly lit café, but his shirt continues to look the same shade of red. The shirt doesn't look any darker despite the fact that there has been a significant change in the wavelength of the light reflected from it onto your retina. How are we to explain this? For Churchland (2012), thinking of perception as grounded in the operations of a simple recurrent network provides a good explanation. How you consciously perceive the colour, argues Churchland, is a matter of the activity at the middle layer of a network concerned with colour perception. How that middle layer responds to the light input when you enter the café will be influenced by how you perceived the shirt just before, when you were outside. This is because the pattern of activity at the middle level at that earlier point in time is stored in the context layer and fed back into the network alongside the current activity at the input layer.

5 Evaluating connectionism

How powerful a challenge is connectionism to classical computationalism? A first point that needs to be made is that the two approaches are not mutually exclusive. On the one hand, perhaps some of our cognitive capacities are underpinned by connectionist networks, whereas others are underpinned by classical computational mechanisms (Marcus, 2001). Such a hybrid approach might be developed to engage with Churchland's point, discussed above, that classical computationalism is most at home with those aspects of human cognition that we express linguistically. Thus, so it could be argued, what marks humans apart from our non-human relatives is that we evolved classical computational mechanisms to sit on top of phylogenetically older connectionist networks and that those mechanisms are involved in the kind of higher-level cognitive processing that connects with language but not with other more basic cognitive capacities.

On the other hand, connectionism could be married with classical computationalism by viewing classical computational processes in the mind-brain as having been implemented at a lower level by connectionist networks (Fodor and Pylyshsyn, 1988; Hinton et al., 1986). Thus, the classical computationalist and the connectionist would not be in opposition to one another but would be operating at different levels in a Marr-style hierarchy, with the latter operating at a level below Marr's level of representation and algorithm. Such a perspective might be suggested by a cursory look at contemporary artefactual computers of the kind used to write this book. Such computers are classical manipulators of syntactically structured symbols, but such symbols are encoded by means of strings of 1s and 0s that are realized by the states of electronic circuits which can be either 'on' (so representing '1') or 'off' (so representing '0'). Described in such terms, these classical computational symbols don't look so far removed from activation patterns at a given layer of a connectionist network. Moreover, within a contemporary PC or Mac, electronic circuits are arranged to form logic gates, units that take pairs of binary numbers as input and produce a binary number as output. One example of a logic gate is an AND gate which, when fed <1, 1> as input, produces <1> as output and produces <0> in response to all other input combinations. An OR gate, on the other hand, produces the output <1> in response to all possible input combinations apart from <0, 0>.[26] These logic gates can themselves be combined to form more complex networks that carry out complex logical operations. A key point about logic gates is that they can be implemented on a simple connectionist network, implying that a classical computer composed of logic gates can be implemented by a network of simple connectionist networks appropriately harnessed together.

One historically prominent line of argument in support of connectionism is the so-called hundred-step argument that appeals to the relative speed of the brain and a standard digital computer (Feldman and Ballard, 1982). The power of contemporary artefactual computers has a lot to do with their speed; they execute each operation in a fraction of a millisecond. This basic speed means that a program with many individual instructions can be executed serially – that is, one step at a time – in a short period of time. Brains, or, more specifically, neurons, are much slower, and the upshot of this is that, if the brain were a classical computer, the programs it could run could consist of a maximum of only 100 steps or instructions. But this, so the argument concludes, is not plausible, as witnessed by the complexity of standard AI programs, with the consequence that the brain must be a parallel processor to compensate for the relative slow speed of its basic operations.

I don't think that this argument is particularly powerful, as it is based upon the mistaken assumption that classical computationalism is inconsistent with the view that the mind-brain engages in parallel processing. There is nothing to stop the friend of classical computationalism arguing that, whenever we exercise a particular cognitive capacity, there are lots of distinct computational processes going on at the same time. On this view, cognition would be more like a professional restaurant kitchen – where many individuals are involved in the preparation of a given meal by carrying out different culinary tasks at the same time – than a domestic kitchen where one person does everything one task at a time.

A second argument in favour of connectionism is that it better explains the graceful degradation of cognition than does classical computationalism. Suppose that the input to a connectionist network is 'damaged' in the respect that an incomplete pattern is instantiated at the input level. For example, with respect to Cottrell's face-recognition network, this could involve the network being stimulated by a face image that has an element that is blacked or bleached out. Despite this, the network will often be able to process that input and generate the correct output – for example, identify the owner of the face image. Alternatively, suppose the network is damaged in the respect that some of its constituent units or connections are broken. Nevertheless, the network will not completely collapse as a result of this damage. Rather, its input–output performance will approximate that of an undamaged network and will deteriorate gradually in proportion to the extent of the damage. This gradual deterioration of performance in proportion to the extent of damage is known as graceful degradation. However, artefactual computers do not degrade gracefully; any damage to a computer's input or to its internal components tends to have catastrophic results. The mind-brain, on the other hand, does display graceful degradation. For example, a lesion in the brain caused by a stroke or a heavy blow to the head will have some effect on cognition but will not lead to a complete collapse of a person's cognitive capacities. Therefore, so the argument concludes, the mind-brain behaves more like a connectionist network than a classical computer.

Once again, I am not convinced by this argument, as the graceful degradation displayed by the mind-brain could be due to the way in which classical computational processes are implemented at a lower level. For example, if they were implemented by a connectionist network, as described above, then we should expect graceful degradation.[27]

One of the most widely discussed objections to connectionism has been developed by Jerry Fodor and his colleagues, who argue that connectionism cannot account for the systematicity of thought (Fodor and Pylyshyn, 1988; Fodor and McLaughlin, 1990). According to Fodor, thought is systematic in the respect that anyone capable of thinking aRb (that is, that object a stands in relation R to object b) is also capable of thinking bRa, and vice versa. For example, anyone capable of thinking that the dog chased the postman is also capable of thinking that the postman chased the dog, and vice versa. For Fodor it is nomologically necessary that thought is systematic; that is, it is a law of human psychology.

Fodor argues that the systematicity of thought is something that requires explanation and that the classical computationalist can produce such an explanation whereas the connectionist cannot. To recapitulate, the classical computationalist explanation of the systematicity of thought runs as follows. Having a thought involves deploying a language the sentences of which have syntactic structure. Thus thinking aRb involves forming a sentence in the mind-brain that has constituent symbols relating to a, b and R that are put together to form a particular syntactic structure. Thinking bRa involves deploying the very same resources – that is, the same symbols and the same symbol-combining mechanisms – to form a different yet analogous symbol. In short, languages are perfectly designed to be systematic, so that, by employing languages, systematicity falls out of the nature of classical computation.

Fodor objects that, in contrast, connectionism cannot explain systematicity because the key systematicity supporting features of classical computers is not an inevitable feature of connectionist networks. Suppose a connectionist network is capable of thinking or representing aRb. It will do this by means of the activation of a node or a pattern of activation over a number of nodes and not by means of a syntactically structured symbol. Consequently, there is no guarantee that the network is also capable of representing bRa; that is, there is no guarantee that there will be a node or a pattern of activation that means bRa. This is not to say that a connectionist network could not be systematic, as there are networks with this characteristic (Smolensky, 1991; Marcus, 2001), as Fodor is willing to concede. But, he argues, this is beside the point, as systematicity is a nomologically necessary feature of human thought. Thus, what needs to be explained is why it is a law that human thought is systematic, and the connectionist cannot explain this, for it is not part of the nature of connectionist systems to be systematic; if a network is systematic, it is only an accident that it is so.

I'm not convinced by Fodor's argument. I agree that the connectionist doesn't have access to an easy, direct and general explanation of systematicity. However, this doesn't rule out the possibility of a relevant explanation. Suppose one has a systematic network. One could then explain how it manages to be systematic by appeal to its specific features – for example, by appeal to the meaning of its various potential patterns of activation, the connection weights between the constituent units, the threshold values of the units, and so on. Thus, if a connectionist had a substantial, fleshed-out theory of the networks that support human thought, then they could appeal to the specific features of those networks to explain their systematicity. I find it difficult to see why this is not explanation enough.

Another objection to connectionism focuses on its account of learning. Learning by means of back propagation is a lengthy process that requires knowledge of the correct output to any input presented to the network, so that the extent of divergence of the actual output from the correct output can be calculated. But it is hardly plausible that this is how we generally learn, as we are typically not rigorously supervised while we are learning.[28] Connectionists such as Churchland (2012) are often happy to accept this criticism and concede that back propagation does not constitute a biologically plausible model of learning and cognitive development. However, they argue that there are alternative ways for a connectionist network to learn that do not require supervision; for example, Churchland appeals to Hebbian learning, a process named after the Canadian psychologist Donald Hebb (1949). According to Churchland (2012), Hebbian learning within a connectionist network takes the following form: when a number of units in a particular layer of the network are simultaneously active, the connections are strengthened between each of those units and any units in the next layer to which they all connect. For example, suppose that two units, x and y, in the input layer of a network both have excitatory connections to unit z in the middle layer. On one occasion x and y become highly active, simultaneously sending stimuli to z that cause it to become highly active. This will result in the connections between x and z, on the one hand, and y and z, on the other, to strengthen. The upshot of this may well be that, in the future, the activity of x alone (or y alone) will be sufficient to push z into activity.

With respect to the relative strengths of classical computationalism and connectionism, I offer an ecumenical conclusion. Neither approach can be dismissed at this stage in our knowledge of human cognition, and both may well turn out to offer lasting insights. Rather than seeking to identify ourselves in global terms as either classical computationalists or connectionists, we should keep an open mind and seek to identify the best theories currently available for any aspect of cognition that interests us.

6 Alternative approaches to the study of cognition

We have seen how the rise of connectionism constituted a move away from the classical computational view of cognition central to early cognitive science, with its commitment to language-like syntactically structured representations manipulated by means of internal rule-governed processes. The 1990s saw some more radical developments that, according to Andy Clark (2014: 140), continued this 'retreat from the inner symbol'. Indeed, the most radical figures at the forefront of these developments[29] questioned whether representations had any role to play in cognition. In this section I will examine some of these developments. It is worth noting two points at the outset. First, as remarked by Wilson and Clark (2009) and Shapiro (2011), the developments in question, rather than exemplifying a unified approach underpinned by a coherent set of common assumptions and aims, constitute a family of approaches variously labelled 'the situated and embodied cognition approach', 'Artificial Life' and 'Dynamical Systems Theory'. They are also sometimes called the four Es, because they claim that cognition is embodied, embedded, extended (in the sense that cognitive phenomena can extend beyond the skull) and enacted (in the sense that cognition constitutively involves action). Second, although they have come to prominence relatively recently, they draw upon, and in some cases resurrect, much older ideas such as those found in cybernetics (Dupuy, 2000), Gibson's (1979) approach to vision, and the work of continental philosophers such as Heidegger (Wheeler, 2005) and Merleau-Ponty (Gallacher, 2005).

One helpful way of giving an initial sense of the general thrust of this family of approaches involves contrasting them with a Cartesian view of cognition. Cartesianism is a view about the nature of the mind and its relationship to the body and the external world that is associated with the work of the great French philosopher René Descartes. According to Descartes (1985), the physical world is a mechanical, law-governed system occupied by things whose essential characteristic is to be spatially extended.

Human bodies are examples of such inhabitants of the physical world and are, accordingly, spatially extended physical systems. A human individual consists of two components, a body and a mind. Minds are essentially thinking things. The mind and the body causally interact with each other. For example, in perception the body is stimulated by the external world and so provides input to the mind, which then engages in reasoning to build up complex representations about that external world. How the body moves, and so how we behave, is an effect of the mind's activity as, taking into account its knowledge of the world and its desires, the mind plans by reasoning how to behave, issuing instructions to the body that cause it to move in the way that it does.

Descartes thought that the mind was a non-physical thing and so he was a substance dualist, but in the relevant respect it is possible to be a materialist Cartesian. Early computational cognitive scientists were indeed materialist Cartesians. For them, the mind is the brain, and the mind-brain engages in reasoning activity involving internal representations that encode the individual's knowledge. The mind-brain is separate from the body and the external world but causally linked to the former, and this causal relationship allows it to interact with the world. Early computational cognitive scientists also viewed perception as a matter of the mind's building up representations on the basis of input coming from the body and action as the effect of processes of planning executed by the mind. In other words, the mind connects with the body and the world only at its 'outer surfaces', so that its internal workings can be studied without much concern with the body or the world.[30]

In the 1990s a number of cognitive scientists began to express their reservations about this Cartesian conception of the mind and to develop alternative approaches. Their outlook was underpinned by the following ideas. First, studying how simple systems – for example, insects, rats, simple robots and machines – solve simple problems can shed light on human cognition. Second, a crucial fact about cognitive systems such as humans is that they are embodied and embedded or situated in an environment. Thus, we cannot profitably study cognition without concerning ourselves with our embodiment and our situation in the world. Third, we are products of evolution. As François Jacob (1977) is famous for having pointed out, evolution works by tinkering, so products of evolution may not be perfectly designed from the engineering perspective. Fourth, from an evolutionary perspective, the central feature of cognitive agents is that they engage in purposive behaviour – that is, they behave in such a way as to satisfy their biological needs for food, shelter, sex, and the like.

Motivated by these ideas, some cognitive scientists began to study how simple embodied systems utilizing features of their immediate environment behave in such a way as to fulfil their basic goals. To see the kind of

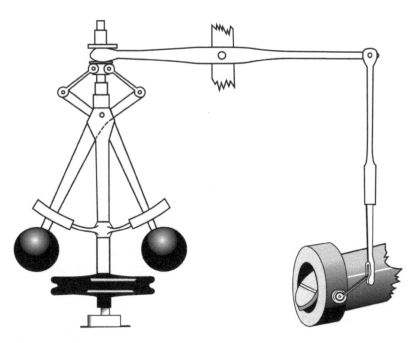

Figure 2.5   The Watt governor

picture of cognition that emerged, consider van Gelder's (1995) classic example of a Watt (centrifugal) governor.

The Watt governor was designed in 1778 by the Scottish engineer James Watt to solve a problem faced when driving industrial machinery by a steam engine (see figure 2.5). A flywheel, which needs to be driven by a belt, is connected to the machinery so that its rotation will drive the machinery. The flywheel rotates as a result of the operation of steam-driven pistons. The delivery of steam to the pistons is controlled by a throttle, so that opening the throttle causes the flywheel to speed up and closing it causes it to slow down. The problem is to control the opening of the throttle so the flywheel rotates at a constant desired speed. How is this to be done? Van Gelder describes a solution to this problem that involves a computer being fed information concerning the current speed of the flywheel, carrying out computations by drawing on representations of the desired speed, the relationships between speed and throttle opening, and so on, and ultimately issuing instructions to a mechanism controlling the opening of the throttle. This process involves a cycle of sensing, modelling (that is, representing), planning and acting. Now, although this is a possible way of solving the problem, says van Gelder, it is not the solution adopted by Watt. Watt's solution was much simpler and relied upon an ongoing process of reciprocal causal interaction between components of the machine that relies heavily on a salient feature of the environment.

Such reciprocal causal interaction is known as coupling. In greater detail, here is how Watt's solution goes. Attached to the spindle of the flywheel is a pair of hinged arms with weighted balls on the end. Due to the action of centrifugal force, as the flywheel rotates the arms rise; the faster the rotation, the more the arms rise. The arms are in turn connected to the throttle, so that raising them has the effect of closing the throttle and lowering them has the effect of opening the throttle. Consequently, as the speed of the flywheel increases, the arms rise, causing the throttle to close. This in turn causes the flywheel to slow down. As the flywheel slows down the arms drop, which opens the throttle so that the flywheel speeds up again. If appropriately calibrated, this ongoing two-way process of causal interaction between the arms and the throttle can ensure that the wheel rotates at a near constant desired speed.

Van Gelder's suggestion is that cognition may well resemble the workings of a Watt governor and so not rely upon representations and computations involving them. Rather, the purposive behaviour that is central to being a cognitive agent could be the result of an ongoing process of reciprocal causal interaction between the body and the brain that relies upon key features of the environment. In such a case, not only would the brain's activities not involve computation but the brain alone would not be the seat or location of cognition. Rather, as the brain is coupled to the body, the brain and the body together would constitute a unified cognitive system.

One might object that industrial machines are nothing like people, so that insights into how the former operate offer no insights into how the latter operate. However, work in AI and robotics where simple robots have been designed to navigate rooms has suggested considerable parallels between the behaviour of such systems and the Watt governor. For many, this serves to motivate the kind of non-representational approach championed by van Gelder. Particularly significant here is the work of Rodney Brooks (1991), who designed simple robots called animats that utilized a subsumption architecture to enable them successfully to navigate their local environment.

Robots with a subsumption architecture contain subsystems that work independently and in parallel. These subsystems each take sensory input and generate particular behavioural routines that are associated with a specific goal. They are organized into layers, with subsystems associated with higher-level goals being 'layered' on top of subsystems associated with more basic goals. In this way, the exercise of the higher-level goals can draw upon the exercise of the lower-level goals. An early example of a robot with a subsumption architecture was named Allen. Its lowest level layer is concerned with object avoidance and operates so as to cause Allen to halt or to swerve if directly confronted by another object. The next layer is the wanderer subsystem, which is responsible for the robot's random wandering. The highest level layer is the explorer subsystem, which is responsible for Allen's exploring his environment in a systematic manner. Such exploring requires wandering, which in turn requires avoiding obstacles, so each subsystem subsumes those at the lower layers. A crucial feature of the design is that an output delivered at a given layer can impact on a higher-level layer, as when the obstacle-avoidance layer issues a 'halt' instruction when it senses an object directly in front of Allen. The performance of robots with a subsumption architecture far outstripped that of traditional AI creations, such as SHAKEY,[31] that relied upon the sense, model, plan and act cycle.

Brooks notes that representations are structures that act as proxies for the things that they represent and so are useful for performing tasks relating to those things when they are not present. For example, if you have a photograph of Barack Obama, you can reliably answer questions about his appearance when he is absent by consulting the photograph. But if Obama was right in front of you, then you would have no need for the photograph as you could look directly at him. Similarly, as Brooks's robots are engaged in navigating an environment in which they are embedded – that is, one that is directly present to them – they have no need for representations that stand in for that environment; rather, there is a direct link between the environment and the robot's behaviour.

7 Evaluating anti-representationalism

How convincing is this anti-representational perspective? The approach is based upon the crucial assumption that human cognition operates in broadly similar ways to that of simple systems, be they robots, machines or animals.[32] On the face of it, this is not an outlandish assumption, especially when one takes a biological, evolutionary perspective. For one would expect nature to build complex systems that solve complex problems by utilizing the solutions that their simpler ancestors employed to solve simpler problems. However, this line of thought would be put under pressure if we could find a fundamental discontinuity between us and such simpler systems. I think that there is such a discontinuity. Simple systems such as machines, robots and animals behave in fairly fixed and limited ways in order to solve a narrow range of problems that are dictated by their design or biology. On investigation, the solutions to such problems often turn out to be less intelligent than first appears. And, typically, they involve responding to readily detectable features of the environment; such environmental features serve to trigger items from a fixed repertoire of behavioural options. The classic example here is the Sphex wasp (Dennett, 1978b). The Sphex wasp lays its eggs in a burrow, and to provide the emerging grubs with food it places a paralysed cricket in the burrow. Its preparations appear intelligent, as it drags the paralysed cricket to the entrance of the burrow and leaves it there while it goes inside the burrow to check that all is in good order. Only then does the Sphex drag the cricket inside. However, this appearance of intelligence evaporates if the cricket is moved a few centimetres away from the entrance of the burrow while the Sphex is inside. What happens then is that the Sphex drags the cricket back to the entrance of the burrow after completing the initial inspection and reinspects the burrow while leaving the cricket at the entrance for a second time. If the cricket is moved away once more, the Sphex repeats the process a third time, and so on ad infinitum.

Looked at this way, we humans seem to be very different. There is a vast array of goals a human individual can in principle adopt that far outstrip our basic biological needs, and these goals can differ substantially from person to person or between different time slices of the same person. How we go about achieving any given goal can vary immensely and is sensitive to the beliefs and desires that we contingently have. And our behaviour is stimulus independent in that there are few features of the environment that compel us to respond in a particular fixed, uniform way. In short, human behaviour and goals exhibit a variety and flexibility unlike that of the systems investigated by champions of anti-representational approaches.[33] It seems to me that such variety and flexibility cries out for explanation in terms of representations; the processes underpinning such variety and flexibility are, to borrow Clark and Toribio's (1994) term, 'representation hungry'.[34] Accordingly, in the remaining chapters of this book, where I discuss a range of cognitive phenomena such as conceptualization, language and perception, representations will loom large. For, whatever the excitement surrounding the emergence of anti-representational approaches, it is just not the case that cognitive science has generally eschewed representations. However, I will return to anti-representational approaches in chapter 6, where I discuss a view of perception known as enactivism.

8 The extended mind

Before closing this chapter, I will look at another recent development that puts pressure on the orthodox idea that cognition is always located within the brain. This is the thesis of the extended mind, according to which cognitive processes can extend beyond an individual's outer surfaces. The landmark work here is the article 'The Extended Mind' by Clark and Chalmers (1998). The authors discuss the case of Otto who, because of the deterioration of his long-term memory, relies upon a notebook containing the information that he needs to go about his daily life. So, for example, if Otto wants to go to the Museum of Modern Art, rather than drawing on his biological memory, he will consult his notebook, which contains information regarding the museum's location. In fact, his notebook is integrated into his life in such a way as to parallel that in which a normal person's bio-logical memory is integrated into their life. The question, then, is whether Otto's notebook is part of his mind, so that the cognitive processing that draws upon the information contained in his notebook literally extends beyond the outer surfaces of his brain and body. Clark and Chalmers argue for a positive answer to this question on the grounds of the parity principle, according to which:

> If, as we confront some task, a part of the world functions as a process which, were it to go on in the head, we would have no hesitation in accepting as part of the cognitive process, then that part of the world is (for that time) part of the cognitive process. (1998: 8)

The idea is that, given that Otto's notebook is integrated into his life in a way that is functionally equivalent to that in which biological memory is integrated into that of a normal person, it would be arbitrary and unjusti-fied to distinguish between the two cases from the perspective of cognitive science.

The thesis of the extended mind has generated much heated discus-sion.[35] With respect to Otto, I'm not convinced that the required functional equivalence is present. Otto utilizes perception to access the contents of his notebook, but the very perceptual mechanisms that he so employs are typically used (both by regular humans and by Otto) to detect features of the external world, a function that they evolved to perform. As Clark and Chalmers would presumably not want to say that a ripe red tomato seen by Otto was located in his mind, why should we say that the notebook or the markings inside it are located in his mind? Clark and Chalmers discuss an objection to their claims that appeals to perception. Their response is that the presence of visual phenomenology should not make us conclude that Otto's consulting his notebook is not a cognitive process located fully in his mind. I agree, but my point is not about phenomenology. Rather, it is that Clark and Chalmers are forced to drive a wedge between regular perceptual episodes and those where Otto is looking at his notebook. Now suppose that Otto sees a ripe red tomato. Light reflected off this object onto his retina will stimulate an internal process involving the generation of representations that not only represent an external object but represent it as being external. Now suppose Otto consults his notebook. As he is using the same perceptual mechanisms that he uses when he sees the tomato, won't he thereby represent the pages of the notebook and the markings on them as being external to him? In other words, in contrast to his relation-ship to his biological memory, from Otto's perspective the notebook will be external to him.

Nevertheless, I'm not opposed to the idea that the mind could in prin-ciple be extended beyond the skull and our outer surfaces. However, for several reasons, I don't think that this concession involves endorsing anything inherently radical. First, it wouldn't involve abandoning classical computationalism so long as the extra-cranial structures and processes were of the right kind, as they seem to be in the case of Otto. Hence, the thesis of the extended mind is not as substantial a challenge to ortho-dox cognitive science as the anti-representational approaches discussed above. Second, the possibility of cognition extending beyond the skull is perfectly consistent with its being the case that the overwhelming major-ity of actual cognitive states and processes are firmly located in the skull. Third, processes internal to the skull would seem to have a priority over those located outside of it. This is because the latter have to become suit-ably harnessed to the former to become part of our cognitive lives, but not vice versa. This can be seen in the case of Otto. Before enlisting the help of a notebook to deal with his failing memory, Otto was able to rely upon his internal biological memory, and so what existed in his skull constituted a unified cognitive system. The notebook became part of a cognitive system only when appropriate links were established between it and the brain, when, so to speak, the brain brought it within its orbit.

It might be objected that I'm overlooking a common case where cognition spreads beyond our outer surfaces, one which is absolutely fundamental to contemporary human cognitive life. This is the use of writ-ten language in thinking (Clark, 2008). The idea here is that when we write down our thoughts we are not merely expressing pre-formed thoughts. Rather, the writing is part of the thinking process, and we think much better for being able to write. A similar point can be made with respect to the use of paper and pencils in carrying out complex mathematical calculations.

I don't deny that writing can aid thinking but wish to resist the idea that the writing is literally part of the thinking process. My perspective here draws upon Ray Jackendoff's (2002, 2012) view of the function of language. Jackendoff argues that the chief function of language is to help make our thoughts available to consciousness. We can have thoughts independently of language but such thoughts are difficult to introspect; they just don't have an introspectable form. Thus, without language we can know what we think only in broadly the same way that we come to know what other people think: by observing our own behaviour. Once we have mastered a language we have an additional means of manifesting our thoughts, and we can do this out loud – when expressing our thoughts to other people – or in our heads when speaking internally. For example, I typically find out what I think about a particular topic when I hear myself (either out loud or in my head) answering a question concerning what I think about the topic. The linguistic expression of thought is typically direct and immediate and does not involve first introspecting my thoughts and then using language to describe what I so introspect. Speaking internally, then, is not thinking itself but an internal manifestation of thinking that is readily introspectable or present to consciousness. By providing us with a means of internally manifesting our thoughts, language aids thinking considerably. For we can indirectly bring our thoughts to consciousness by being directly aware of their linguistic manifestation. This enables us to hold onto a thought for more than a fleeting moment, as a record of its linguistic manifestation is held in working memory. This in turn facilitates the examination of the nature, grounds and implications of the thought.

Now let us turn to writing. Working memory is notorious for its limitations with respect to how much information it can hold and how long it can hold any given item of information.[36] Mastery of writing can help an individual overcome these limitations as she can commit her thoughts to paper. This provides a record of what she has been thinking and enables her to continue a train of thought without running the risk of irretrievably losing the earlier thoughts. In short, just like internalized spoken language, written language enables an individual to manifest her thoughts in such a way as to make them accessible to consciousness. Moreover, it has the advantage of extending the time frame and bandwidth of that conscious access, so enhancing the individual's capacity to examine the nature, grounds and implications of her thoughts. This doesn't imply, however, that the written words constitute part of the individual's mind. This is because they are accessed perceptually, with the upshot that they aid the conscious recall of previously entertained thoughts by prompting the individual to revocalize their thoughts internally, a process that draws upon the individual's knowledge of language.

9 Conclusion

In this chapter I have examined the dominant general theories of cognition within cognitive science. The two mainstream alternatives are classical computationalism and connectionism. Both of these are versions of the computational theory of mind and, accordingly, postulate inner representations to explain cognition. Moreover, both constitute viable options as far as we can currently see and need not be mortal enemies. Anti-representational approaches constitute an alternative to both classical computationalism and connectionism, yet, despite the grand claims made by some of their champions, they currently exist on the fringes of the mainstream and are somewhat unproven and underdeveloped.

3 Modularity

1 Introduction

The focus of this chapter is a heated debate about the architecture of the mind that centres upon the postulation of mental modules. To be an advocate of the modularity thesis is to hold that the mind contains a number of task-specific subsystems that operate in relative independence of one another. In cognitive science, the modularity thesis came to prominence in 1983 with the publication of Jerry Fodor's book *The Modularity of Mind*. Here Fodor argued that there are a number of modular input systems existing alongside a non-modular central processing system. Beginning in the early 1990s, several evolutionary psychologists and their philosophical allies sought to go further than Fodor by arguing for the massive modularity thesis. According to the massive modularity thesis, the mind consists of hundreds – if not thousands – of modules, leaving little space for a domain-general central processing system. Moreover, these modules are a product of evolution by natural selection, having developed to deal with the problems our ancestors faced in the Pleistocene epoch (a period of geological time stretching from approximately 2,600,000 to 11,700 years ago). In this chapter I will examine the plausibility of the various versions of the modularity thesis.

2 Fodorian modules

In the introduction to this chapter I characterized a module as a relatively independent task-specific subsystem of the mind. On reflection, such a characterization is a little too weak, as few cognitive scientists would dispute the claim that the mind contains modules of this kind (Samuels, 2006). Thus we are faced with the question as to what exactly a mental module is. Cognitive scientists have offered different answers to this question. Fodor (2000, 2001) represents Chomsky as thinking of modules as bodies of knowledge. Thus, the language module[1] is a body of knowledge about language drawn upon in linguistic processing.[2] Fodor (1983) himself – though resisting providing a definition of the term 'module' – identifies nine properties of perceptual systems and argues that, if any psychological system has all or most of these properties, then it counts as a module. However, he places particular emphasis on two of them, namely those of being domain specific and being informationally encapsulated. Domain specificity is a matter of having a specific subject matter, which in turn involves processing only a narrow range of inputs.[3] A subsystem is informationally encapsulated when there is information stored in the broader system of which it is a part to which it does not have access even when that information is relevant to the tasks that it performs. Such a subsystem is forced to rely upon its own specialist store of information. Thus, we can say that a subsystem of the mind is a Fodorian module if and only if it is domain specific and informationally encapsulated.

Although Fodor has been described as operating with an overly strong notion of modularity[4] understood in this way, he presents a workable general notion of what a module is. This is because those who have been critical of his understanding of modularity tend to object to the attribution of one or more of the other seven properties to modules. For example, Carruthers (2006a) objects to the idea that modules typically have shallow outputs (that is, outputs that fall short of making a judgement or endorsing a belief) or that their operation is fast. And Karmiloff-Smith (1996) objects to the idea that modules must draw upon innate information.

Just as advocates of the modularity thesis can disagree with respect to the question of what a module is, they can also disagree with respect to the question of how many modules there are and the identity of those modules. On this front, Fodor is relatively parsimonious. He distinguishes between three distinct types of subsystem that exist in the mind. These are the central system, input and output systems, and transducers. The central system is the home of propositional attitudes, such as beliefs, desires, and the like, and executes reasoning processes involving such states. Transducers lie at the interface between the mind and the world. Input transducers are stimulated by the external world and produce representations in response to such stimulation. For example, the retina is an input transducer which is stimulated by light reflected from external objects and produces representations of the wavelength and intensity of that light as its output. Output transducers operate in the opposite direction, transforming symbolic input into non-symbolic output. For example, motor transducers involved in the initiation and control of action transform instructions to produce particular bodily movements into neural impulses that cause the relevant muscle contractions needed to bring about such movements. Input and output systems lie between the central system and transducers. Input systems take the output produced by an associated transducer as their input and then perform computations on it so as to generate a representation of the distal cause of that input. That output representation is then passed to the central system, which decides on that basis what beliefs to form about the external world. Output systems, on the other hand, take as their input from the central system representations relating to decisions to execute particular actions. They then compute what specific bodily movements need to be made to perform these actions, passing the results of this processing to associated output transducers in the form of instructions to produce particular bodily movements.

Fodor claims that input and output systems are modules, though after making this claim he focuses his attention on input systems, asserting that there are six such modules in the human mind. Five of these modules relate to the five senses as traditionally conceived (that is, seeing, hearing, touching, tasting and smelling). The sixth is the language module responsible for linguistic processing. However, Fodor is insistent that the central system does not have a modular structure.

What reasons are there for agreeing with Fodor that input systems involved in perception and linguistic processing are modules? One prominent argument produced by Fodor relates to the existence of perceptual illusions. Consider figure 3.1.

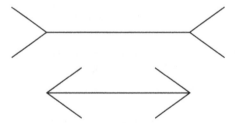

Figure 3.1

Which of the two lines looks the longer? To a typical human perceiver the upper line looks longer than the lower line. But in actual fact the lines are of exactly the same length. This is known as the Müller–Lyer illusion. What is interesting about this illusion is that it persists even when the perceiver knows that the two lines are of the same length. What this suggests is that visual perception is informationally encapsulated in that there is information stored in the mind concerning the relative length of the lines to which the visual system does not have access even though it is highly relevant to the task at hand. The informational encapsulation of vision, along with the seemingly obvious fact that the visual system is sensitive only to light, implies that the visual system is a module. There are other visual illusions that have the same implication – for example, that produced by the Ames room and the Herring illusions.[5]

At this stage it would be a little too quick to conclude that vision and other perceptual systems are modular. Recent work on vision suggests that visual processing is not purely bottom-up. That is to say, the visual system does not take its input and then process it inexorably so that the same input will always result in the same output. Rather, visual processing has top-down elements in the respect that relatively high-level information can influence the way that lower-level information is processed. One example of this relates to the perception of colour. One might expect that the colour that one perceives a surface as having is exhaustively determined by the wavelength of the light reflected off it onto one's retina. However, the phenomenon of colour constancy indicates that this is not the case. The lighting conditions that we confront change considerably from one moment to the next, as we move from one illuminated room to another or from inside to outside, and as the sun sets or is obscured by moving clouds. These changes impact considerably on how the retina is stimulated. Nevertheless, the colours of the surfaces we see do not routinely appear to change throughout these developments. For example, the colour of your companion's shirt will not appear to you to change as you move from inside a dimly lit café to a sun-drenched street. In other words, the colours of the surfaces that we see remain constant across a range of changing lighting conditions.

What seems to support colour constancy is an assumption built into the visual system that surfaces typically maintain their colour from one minute to the next and that changes in lighting conditions can bring about changes in the nature of the light reflected off any given surface. Moreover, the visual system has 'knowledge' of the systematic relationship between lighting conditions and the quality of light reflected off surfaces and can access information about the prevailing lighting conditions in order to enable it to compensate for any anomalies, so that surfaces appear to the individual to have the colour that they actually have.

In and of itself the existence of top-down visual processing does nothing to undermine Fodor's claim that the visual system is a module, so long as the relevant higher-level information is stored in the visual system and not drawn from information in the central system that can be utilized for any number of cognitive tasks. However, in the cognitive psychological and cognitive neuroscientific literature there is a tendency to talk of an individual's propositional attitudes as influencing her visual experiences.[6] On the face of it, this seems to cause Fodor problems. But, on the other hand, I detect a looseness in the use of propositional attitude terms in such contexts, so that the writer could better be interpreted as referring to subpersonal states internal to our perceptual systems. For example, Smith and Kosslyn write:

> Studies of visual illusions have revealed that context - including our knowledge, beliefs, goals and expectations – leads to a number of different

assumptions about visual features. ... [For example] ... *size illusions* demonstrate that we assume objects maintain their true size across changes in apparent distance from the observer. (2007: 85)

The problem with this is that it is more natural to say that it is, first and foremost, the visual system which is doing the assuming and that these assumptions reside at the subpersonal level.

A more plausible case of propositional attitudes entering into the visual process and influencing how we subsequently see the external world relates to attention. Attention is the process whereby certain information is attended to so as to receive further processing, whereas other information is either ignored or suppressed. Whenever the visual system is stimulated there is much information implicit in that input. In fact, there is far too much implicit information contained in the visual input for it all to be brought out by further processing without overloading the visual system. Consequently, the visual system must select which elements of its input to process and which to ignore. Psychological experimentation has revealed some surprising results of this process of selective attention. For example, in a famous experiment, subjects intently watching a basketball game, under the instruction of counting the number of passes made by players wearing white, failed to notice a woman in a gorilla suit walking right in front of them (Simons and Chabris, 1999). The potential problem for Fodor here is that it would seem that propositional attitudes can influence what we attend to. Smith and Kosslyn (2007) produce an example of an individual at a loud and busy party looking for a friend he believes to be wearing a green dress. Wanting to locate his friend and having a true belief about what she is wearing enables the individual to locate her quickly by influencing which elements of the visual input are fully processed. However, it can be argued on Fodor's behalf that such cases of attention don't imply that vision is not informationally encapsulated. This is because we need to draw a distinction between two distinct ways in which visual processing might be influenced by an individual's propositional attitudes. The first involves the visual system's having a task set for it by an individual's propositional attitudes. If the task is 'find Sarah who is wearing a green dress', this might result in the visual system's scanning the scene and engaging in any further processing only when a person wearing green is spotted, so that a lot of visual information is selectively ignored or cast aside. The second involves drawing upon information encoded in an individual's propositional attitudes when visually processing any element of visual input. Now the work on attention that I have discussed can naturally be interpreted as being an instance of the former kind of process. But to be problematic for Fodor it would have to be an instance of the latter kind of process. In short, Fodor is not committed to the claim that there is no way in which propositional attitudes can influence perception.

My reflections on attention provide a ready response to Prinz's (2006a) attempt to undermine the claim that perception is modular by appeal to top-down processing. Prinz argues that, in the case of ambiguous images such as the duck-rabbit (which can be perceived either as a duck or as a rabbit; figure 3.2), verbal cueing can influence how the image is experienced. And in the case of the Necker cube (which can be perceived either as a left-facing cube or a right-facing cube; figure 3.3) we can choose how to experience the image.

Figure 3.2   The Duck-Rabbit figure

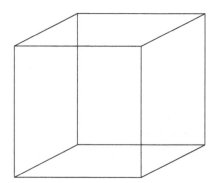

Figure 3.3   The Necker Cube

Prinz takes these phenomena to be inconsistent with the modularity thesis, as they are cases where beliefs impact upon perceptual processing. My objection is that, in such cases, beliefs do not play the right kind of role to support Prinz's conclusion. This is because they merely set the visual system a particular task (for example, try to see the picture as a rabbit) which it is capable of performing independently of that belief. Or, alternatively, they direct it to producing a particular output (say, that of representing the image as being of a right-facing cube) which constitutes one of only a small number of outputs that the visual system's internal structure allows. What Prinz needs to show is that beliefs are drawn upon in the course of processing, and this is what he hasn't done.[7]

Recent work on the perception of colour might not be so easy to accommodate within a modular perspective. In an experiment conducted by Hansen et al. (2006) subjects were shown a digital photograph of a banana that had been adjusted so that the image was grey in colour. Nevertheless, the image was experienced as being yellow. The interpretation of this result is that the individual's knowledge that bananas are generally yellow influenced the processing of her visual system so she came to see it as yellow. If this phenomenon is endemic, then it would be problematic for Fodor, in implying that our beliefs about the colour of familiar objects are systematically drawn on in visual processing. Perhaps the best way of attempting to deal with this problem is to point out that Fodor does think that the visual system engages in basic object categorization. The degree of categorization relates to basic categories. An example of a basic category would be that of *dog* but not the more general *animal* or the more specific *Labrador*. This implies that information about basic categories (perhaps including their typical colour) is stored in the visual system so that it is drawn from that store when used in visual processing rather than a general store of beliefs. With respect to the banana example, the suggestion would be that *banana* is a basic category, so that the visual system is likely to store information about the colour of bananas. For this response to work, it needs to be established that the phenomena revealed by the experiment are restricted to a relatively small class of basic categories. And, for that to be done, further research is needed.

A second argument for the claim that input systems are modules relates to evidence that some of our capacities disassociate from one another. If input systems are modules, then the cognitive capacities that they support will be independent of one another and of capacities associated with the central system such as general intelligence. As a consequence it will, in principle, be possible for an individual to have a specific impairment with respect to a capacity associated with a particular input system while their general cognitive capacities remained intact. Such an impairment would be the result of damage to a particular module caused by genetic or environmental factors. Moreover, it will be possible for an individual to suffer from wide-ranging cognitive deficits due to damage to their central system, as manifested in low general intelligence and difficulty in performing everyday tasks, while being perfectly normal with respect to a capacity associated with a particular input module. In other words, the modularity thesis implies that perceptual and linguistic capacities doubly dissociate from general intelligence. Hence, evidence that these capacities do doubly dissociate would be evidence in support of the modularity thesis.

There is evidence generated by the fields of cognitive neuropsychology and developmental psychology that input capacities and general intelligence do indeed doubly dissociate. The cases most widely discussed in the literature follow Pinker (1994) in focusing on language. Williams syndrome (WS) is a rare genetic disorder. Subjects of WS typically have 'elfin-like' faces, are prone to hypertension, have a below average birth weight and poor muscle tone, and suffer from premature ageing of the skin (Karmiloff-Smith et al., 1995). Their general intelligence is severely impaired, as they typically have an IQ in the 50 to 60 range and find everyday tasks, such as tying their shoelaces, finding their way around, and doing simple arithmetic, very difficult (Pinker, 1994). However, there is evidence that, despite their general cognitive problems, WS subjects have well-developed linguistic capacities. For example, Karmiloff-Smith and her colleagues write: 'Compared to their marked difficulties in simple arithmetic, some WS subjects are almost at ceiling on certain tasks measuring understanding of syntactically complex structures. . . . Also, in contrast with autism and Down syndrome, they produce complex narratives that make extensive use of prosody' (1995: 200).[8]

However, as Cowie (2008) reports, not all researchers of cognitive deficits endorse this view of the linguistic capacities of WS subjects, and she lists a number of linguistic abnormalities that have been attributed to them. These include delayed acquisition of vocabulary; a tendency to overgeneralize regular plural and past-tense endings; difficulties with relative clause constructions, with embedded sentences and (for French speakers) in assigning grammatical gender; and poor performance in sentence repetition tasks. Cowie (2008: 52–3) concludes that:

> Findings such as these lead experts such as Annette Karmiloff-Smith to urge 'dethroning the myth' of WS 'intact' syntactic abilities . . . and move Ursula Bellugi – formerly a proponent of the 'spared language' viewpoint – to caution that 'because their language abilities are often at a level that is higher than their overall cognitive abilities, individuals with WS might be perceived to be more capable than they really are.'

I am hardly in a position to adjudicate this debate as to the linguistic capacities of WS subjects, but something might be said on behalf of those who wish to cite WS in support of the view that there is a language module. This view is closely associated with a Chomskyan perspective on language. For the Chomskyan, there is an important distinction between linguistic competence and performance (Chomsky, 1965). An individual's observable linguistic behaviour – including what sentences they produce and how they understand the utterances of their fellows – is influenced by a wide range of factors quite apart from the state of their language faculty. Now the kind of linguistic problems that Cowie lists seem to be precisely the kind that, from a Chomskyan perspective, could be caused by factors that lie outside of the domain of linguistic competence or the language faculty – factors such as memory limitations and poor general intelligence. In short, then, WS subjects do not need to match the linguistic performance of individuals of normal intelligence to support the case for the modularity of language.

A second case of substantial linguistic capacities in the face of general cognitive deficits is provided by Christopher, a man described as a 'linguistic savant' by Smith and Tsimpli (1995), who have conducted a long-running study of him.[9] As a result of brain damage, Christopher performs poorly in non-verbal intelligence tests, is unable to look after himself and has poor hand–eye coordination. Nevertheless, 'he can read, write and communicate in any of fifteen to twenty languages' (1995: 1), and his competence with respect to English, his first language, 'is as rich and sophisticated as that of any native speaker' (ibid.: 78).

As for dissociations running in the opposite direction – that is, involving linguistic deficits combined with normal general intelligence – a plausible case is presented by a genetically based disorder known as Specific Language Impairment (SLI). SLI subjects have a tendency to make systematic grammatical mistakes in speech despite the fact that they grasp the associated concepts. For example, Myrna Gopnik (1994) documents how they have problems with the tense system of language despite the fact that they understand the temporal relations expressed by words such as 'yesterday', 'today' and 'tomorrow'. However, as with Williams syndrome, there is some controversy relating to SLI which potentially impacts on its relevance when arguing for the modularity of language. For example, Cowie (2008) documents research suggesting that sufferers of SLI have problems that go beyond the specific syntactic ones highlighted by Gopnik. For example, they have difficulties with comprehension and have a significantly lower than average non-verbal IQ (Vargha-Khadem et al., 1998).

Before concluding this discussion of the claim that input systems are modules, I will consider an objection developed by Prinz (2006a) to the idea that there is a language module, an objection that focuses on the fact that the modularity thesis implies that input systems are as independent from one another as they are from the central system. Drawing upon work by Marslen-Wilson and Tyler (1987), Prinz argues that conceptual knowledge can affect syntax. He makes the case clearly and directly and so is worth quoting at length:

> In one experiment, subjects are given the following story: 'As Phillip was walking back from the shop he saw an old woman trip and fall flat on her face in the street. She seemed unable to get up again.' The story then continues with one of two sentence fragments, either 'He ran towards . . .' or 'Running towards . . .' In both fragments, the appropriate next word is 'her', but in the first case that choice is determined *lexically* by the prior pronoun in the sentence ('he') and in the second case that choice

is determined *conceptually* (we know that people cannot run when they are lying down). Remarkably, subjects are primed to use the word 'her' equally fast in both conditions. If lexical processing were encapsulated from conceptual processing, one would expect lexically determined word choices to arise faster. These results imply that formal aspects of language are under immediate and constant influence of general world knowledge. (2006a: 31–2)

The problem with this argument is its assumption that advocates of a modularity approach to language draw a rigid distinction between syntax and semantics and regard them as being associated with different modules. A key development in recent linguistics in the Chomskyan generative tradition is something that is known as theta theory.[10] The basic idea is that the lexical entry for verbs contains information about the nature of things that can occupy their subject and object positions. For example, the lexical entry for 'believe' would specify that its subject must be an animate thing. This is why the structure 'the table believed him' is problematic. With respect to 'run', it is plausible that the lexical entry for this verb implies that its subject cannot be lying down. Consequently, all the information that needs to be drawn upon in completing the sentence 'Running towards . . .' is present in the lexical entry for 'run'. Such lexical information is contained in the syntactic module and so does not need to be drawn from an independent store of real-world knowledge. The upshot of all this is that, once one takes into account contemporary views of the language module that utilize theta theory, it becomes clear that a modular view of language predicts Marslen-Wilson and Tyler's experimental data.

This completes my discussion of the claim that input systems concerned with perception and language are modules. In conclusion, I am sympathetic to this claim but accept that it has not been decisively established. I will now turn my attention to the central system, the domain of beliefs, desires, and the like. Is Fodor right to reject the claim that it does not have a modular architecture?

3 Modularity and the central system

Although he thinks that input systems are modules, Fodor is clear that the central system does not have a modular architecture. That is, it does not decompose into domain-specific and informationally encapsulated subsystems. His argument for the domain generality of the central system runs thus. Typically, when we form a belief on the basis of perception, we are required to combine information from several different input systems and take into account our background beliefs. For example, suppose that on a trip to the Buckinghamshire countryside I see a large bird swoop down upon a dead animal. I conclude that the bird is a red kite on the basis of what I see and hear and my background beliefs that red kites are carrion feeders which have become common in rural Buckinghamshire after a successful reintroduction programme in the 1990s. As I incorporate a wide range of information to acquire a belief, the belief fixation process is the opposite of domain specific.

With respect to the non-encapsulation of the central system, Fodor's argument runs as follows. For Fodor, belief fixation is typically a process of non-demonstrative (that is, non-deductive) inference, and such a process involves framing hypotheses and then seeking to confirm them by considering data that bears upon their truth-value. Hypothesis confirmation has two closely related properties that entail that it is not informationally encapsulated. The properties in question are those of being isotropic and being Quinean. Fodor's account of these properties, along with his argument that hypothesis confirmation is isotropic and Quinean, involves an appeal to confirmation in science. Scientific confirmation is Quinean in the respect that, when testing a hypothesis, the hypothesis itself does not tell one how it is to be tested or what observable evidence is relevant to its evaluation. Rather, it is whole collections of theories that have observational implications. Thus, scientists who differ in their theoretical commitments might differ in the significance they bestow on any given observation with respect to a particular hypothesis under test. Scientific confirmation is isotropic in the respect that scientists regard the world as a complex causal system and so do not delimit what information is potentially relevant when testing a particular hypothesis.[11]

Fodor argues that, because it is Quinean and isotropic, scientific confirmation is not informationally encapsulated. For, when deciding whether or not to endorse a particular hypothesis, the scientist doesn't consider just a limited body of information delivered by her input systems but takes into account the entirety of her epistemic commitments. In other words, scientific confirmation is a holistic process. As, Fodor continues, scientific confirmation is a paradigmatic central system process, it can be inferred that all non-demonstrative processes executed within the central system are unencapsulated and so not modular.

In a later work, Fodor (2000) once again argues that central-system processing is non-encapsulated on the basis of an examination of scientific reasoning. Here he argues that, when scientists choose between competing hypotheses, they do so on the basis of simplicity and conservatism. Simplicity is not an intrinsic property of a hypothesis but is relative to the belief system of the scientist in question. Suppose a scientist has to choose between two hypotheses, H and H*. Her belief system might be such that endorsing H would serve to complicate her belief system while endorsing H* would serve to simplify it. One of her colleagues, however, has a belief system such that endorsing H* would serve to complicate her belief system and endorsing H would serve to simplify it. Here is a simple example to illustrate this point. Suppose that H is the hypothesis that Attention Deficit Hyperactivity Disorder is caused by genetic factors and H* the hypothesis that it is caused by dietary factors. The first scientist believes that a wide range of conditions, including hay fever, asthma, eczema, ME, and so on, are caused by dietary factors, and she generally finds little place for specific genes in the etiology of medical conditions. Hence, endorsing H* would bring her view of the etiology of ADHD largely into line with that of her view of the etiology of a whole range of other conditions, whereas endorsing H would make an exception of ADHD. Although there would be no logical inconsistency in thinking that ADHD was quite unlike most other conditions, doing so would serve to complicate her belief system. The second scientist, on the other hand, finds little place for dietary factors in causing medical conditions such as hay fever, etc. Rather, she explains these conditions in terms of specific genetic factors. Consequently, for her, endorsing H in preference to H* would serve to simplify her belief system. What this example shows is that, when determining which of two competing hypotheses is the simpler, a scientist has to consider the entire contents of her belief system.

With respect to conservatism, the second factor involved in choosing between competing hypotheses, the same moral emerges. In the normal course of things, when we endorse a hypothesis and so acquire a new belief, that belief will be inconsistent with some other belief we already hold. To avoid being irrational we will have to update our belief system so that all its elements cohere with one another. The principle of conservatism says that, when choosing between two competing hypotheses, the scientist should, all else equal, choose that which requires the fewest adjustments to her prior held beliefs in order to maintain the general coherence of her belief system. For example, for the typical contemporary biologist, a belief in the truth of Darwin's theory of evolution by natural selection will be central to their belief system, whereas many other beliefs they hold (for example, as to the capitals of various US states or the home grounds of various English lower-league football teams) will be relatively peripheral. But, Fodor points out, this is not merely a matter of the number of adjustments required. Some beliefs are more central to an individual's belief system than others, so that endorsing a new belief that required abandoning a central belief would be a less conservative move than endorsing a competitor that required abandoning only a peripheral belief. The upshot of this is that determining which of two competing hypotheses is the more conservative relative to one's belief system involves working out the relative centrality of any beliefs that one needs to abandon in endorsing the hypotheses in question. But determining how central a given belief is involves examining the whole of one's belief system, and so executing a global and holistic process rather than an encapsulated process.

The moral that Fodor draws from these considerations is not only that the central system does not decompose into modules. In addition he expresses scepticism that cognitive science will be able to shed any light on the nature of central-system processing. In a memorable passage he writes:

> The fact is that . . . global systems are per se bad domains for computational models, at least of the sort that cognitive scientists are accustomed to employ. The condition for successful science (in physics, by the way, as well as psychology) is that nature should have joints to carve it at: relatively simple subsystems which can be artificially isolated and which behave, in isolation, in something like the way that they behave *in situ*. Modules satisfy this condition; Quinean/isotropic-wholistic-systems by definition do not. If, as I have supposed, the central cognitive processes are nonmodular, that is very bad news for cognitive science. (1983: 128)

Indeed (in Fodor, 2000), he goes on explicitly to doubt that central-system processing (in contrast to the processing executed by modular input systems) is computational. This is because computers are very good at encapsulated processing but find global and holistic processing completely intractable.

4 The massive modularity thesis

There has been much heated debate concerning Fodor's claim that the mind contains a non-modular central system. Particularly important are the arguments of a number of evolutionary psychologists and their philosophical allies who endorse what is known as the massive modularity thesis.[12] Their basic idea is that what Fodor regards as the central system is composed of many (perhaps hundreds or even thousands) distinct modules. Consequently, either there isn't a central system or such a system is very limited in its scope and role in human cognition. These modules are computational systems that are the product of a process of evolution by natural selection. Hence, they are sometimes called Darwinian modules (Bermúdez, 2010), having evolved in the Pleistocene epoch in order to solve problems that our hunter-gatherer ancestors faced. There are a number of prominent arguments for the massive modularity thesis that I shall now examine.

The first argument appeals to general biological considerations. As Tooby and Cosmides point out (1992), the human mind, given its relationship to the brain, is a biological system, and so we should expect it to be the product of a Darwinian process of evolution by natural selection. Such an evolutionary process generally produces complex systems with task-specific components that emerged gradually and were able to confer a survival advantage on their host at every stage in their historical development. Hence, we should expect the mind to be made up of task-specific components that conferred a survival advantage on our ancestors in the period when they were evolving. To think otherwise would be to abandon an evolutionary perspective. Peter Carruthers (2006b) develops a specific version of this argument that appeals to the work of the AI pioneer Herbert Simon. Simon (1962) pointed out that the evolutionary process produces complex systems made up of hierarchically organized components, each with a distinctive function within that hierarchy. Carruthers argues that the massive modularity thesis fits this picture, so we should expect evolution to have generated minds with such an architecture.

As Samuels (2006) notes, evolutionary considerations do not serve to privilege the massive modularity thesis over its main competitors. A computational theory of the mind that doesn't postulate a multiplicity of modules in place of a central system does not thereby represent the mind as lacking distinct components. For it is likely to portray the central system as being made up of a number of functionally distinct units, including a working memory, a long-term memory, an executive control unit, mechanisms for executing basic computational operations retrieving items from memory, comparing items, and so on. Indeed, several years before the rise of the massive modularity thesis there emerged a prominent view of psychological explanation that characterized psychological explanation within the orthodox computational tradition as being a matter of decomposing the mind into hierarchically organized components, each performing a distinctive function.[13]

A second argument for the massive modularity thesis involves turning Fodor's argument that central-system processing is global and holistic, and so computationally intractable, on its head. Once again, Carruthers (2006a, 2006b) is a key champion of this argument, which can be described in the following terms. The idea that cognition is computation is the best idea we have about the workings of the mind. Given that computers find global and holistic processes intractable, we are faced with a dilemma: either we abandon the notion that cognition is computation, so that we are left with no idea as to how we cognize, or we postulate a multiplicity of modules in place of the central system. Advocates of the massive modularity thesis take the second horn of this dilemma, on the grounds that a failure to do so makes our cognitive capacities a mystery.

I'm not convinced by this argument, as I think that there are potential ways for a computer to deal with a large and wide-ranging information store so as to make its tasks tractable. To see this, let's return to Fodor's account of scientific confirmation. Recall that Fodor argues that, when the scientist confirms a hypothesis, she takes into account the entirety of her belief system and that scientific confirmation is typical of belief fixation. I think that both these claims are open to challenge. I do not deny that scientific confirmation is Quinean or isotropic, but this doesn't imply that the scientist has to take everything that she believes into account when evaluating a particular hypothesis. Rather, she will operate with a tacit understanding of what information is potentially relevant and what is not, and this will serve to constrain which elements of her belief system she takes into account. For example, it is hardly plausible that a cognitive scientist deciding whether or not to endorse the massive modularity thesis will take into account her beliefs about fundamental physics, geology or the workings of the economy, let alone those about Italian neo-realist cinema, last weekend's football results or the capitals of the states of the USA.

This raises the question of how an individual scientist acquires the tacit understanding that helps her decide what information to take into account and what to ignore or of how she makes such decisions. In considering this question it is important not to lose sight of the fact that scientists do not normally work in isolation but belong to communities, the members of which are engaged in an ongoing dialogue that shapes the understanding of any given community member as to what information is likely to be relevant to what issue. In short, to use Sterelny's (2003) terminology, the scientific community provides the individual scientist with the scaffolding that helps bear the load of the confirmation process. This social scaffolding means that the individual need not engage with considerations that, in principle, might be relevant when seeking to make rational and informed decisions. The upshot of this is that, though it might be true of the scientific community as a whole that it executes a global and holistic process, it needn't be the case that any individual member of the community executes a cognitive process that is global and holistic.

I would also question the claim that scientific reasoning is typical of cognitive processes that result in the acquisition of a belief. Scientific reasoning is a slow, self-conscious, demanding and highly intellectual process and is engaged in by only a small proportion of the human population. Scientists are typically unusually curious and highly intelligent recipients of a lengthy period of specialist training. In all these respects, scientific reasoning differs from many familiar central-system processes. Most people, regardless of their intelligence or educational background, are capable of effortlessly and unselfconsciously categorizing objects, recognizing faces, attributing mental states to their fellows, engaging in practical reasoning, acquiring spatial information and navigating their home environments, making moral judgements, and appreciating music and other art forms. Given these apparent differences between such processes and scientific reasoning, I would advise caution in making general claims about central-system processing on the basis of an examination of scientific reasoning.

In the case of scientific reasoning, I have argued that there is social scaffolding in place to help the individual. But what about all those types of reasoning listed in the previous paragraph where we routinely reach speedy conclusions without the obvious guidance of an external community? Even if, as I have argued, executing such processes need not involve examining the whole of one's belief system, there is still the worry that they may well be computationally intractable in the absence of modules. But perhaps we shouldn't be so sceptical, as the problem may well be that we are operating with an unrealistic notion of what is involved in effective reasoning that is inherited from formal logic. An idea that has become prominent in cognitive science is that we employ heuristics – that is, simple rules that allow us to take short cuts when addressing problems. Such heuristics are frugal in that they do not demand the processing of large amounts of information (Gigerenzer, 2008, 2014). Nevertheless, they are highly effective. Gigerenzer and his colleagues (1999) provide a widely cited example of a non-German individual deciding that Hamburg was a larger city than Mainz on the basis that they had heard of the former but not the latter. Answering the question of which is the larger of two German cities in this way involves adopting a highly effective strategy that gets at the truth without resorting to more traditionally 'logical' and computationally demanding methods. In short, then, it is perfectly possible to make sense of how the central system can solve problems without postulating a multiplicity of modules.

A third argument for the massive modularity thesis draws upon psychological evidence concerning the limitations and strengths of human reasoning and brings us to the influential work of Tooby and Cosmides (1992) and their postulation of a cheater detection module. One historically popular view is that humans engage in reasoning by employing logical rules that they can employ to reason about any subject matter they can think about. So, for example, employing the rule *modus ponens* enables us to conclude that 'Alf eats termites' from the premises 'if Alf is an aardvark then he eats termites' and 'Alf is an aardvark'. However, in the 1970s experimental work by the psychologists Wason and Johnson-Laird (1972) revealed that human subjects have substantial problems when faced with certain reasoning tasks. In the so-called Wason selection task, subjects are shown four cards and told that each has a letter on one side and a number on the other. For example:

They are then asked which card they have to turn over to work out whether the following conditional (i.e., if . . . then statement) is true:

If a card has a vowel on one side then it has an even number on the other.

Most subjects correctly answer that the 'E' card has to be turned over and that the 'C' card does not. However, they incorrectly say that the '4' card has to be turned over whereas the '5' card does not. Subsequent research conducted by Griggs and Cox (1982) in the 1980s has shown that people perform much better in versions of the Wason selection task where the conditional has a different subject matter and is provided with an appropriate context. For example, subjects are told that the following cards relate to individuals drinking in a bar, with one side specifying what a particular customer is drinking and the other specifying their age:

| **Beer** | **Coke** | **25** | **16** |

They are told to imagine that they are police officers checking whether there is any under-age drinking going on in the bar. They are then asked which of the cards have to be turned over to determine whether the following conditional is true:

> If a person is drinking beer then they are over eighteen.

In this test subjects fared much better, as most correctly concluded that the 'Beer' and '16' cards need to be turned over. Yet this is despite the fact that the problem is logically just the same as the earlier one. But this is rather puzzling: why should the subject matter and context of the task make it so much easier to perform? If we always reason by employing domain-general reasoning mechanisms and logical rules, then the subject matter should make no difference.

This is where Tooby and Cosmides (1992) come in, for they have an explanation that appeals to modules, in particular a specific module they call the cheater detector. Cooperation can be very beneficial to human individuals. If we agree that 'if I scratch your back you will scratch mine', then we can both benefit. Yet it will suit your interests better if, when I scratch your back, you don't reciprocate. If I am to benefit from cooperation, I must be able to detect cheaters and freeloaders so that I can avoid being conned by them in the future. Tooby and Cosmides hypothesize that we evolved a module specifically to detect cheaters and freeloaders, as such a module would clearly provide us with a survival advantage. In the second version of the Wason selection task, the subject matter relates to potential cheating and the violation of rules and obligations, and so the cheater detector kicks in, explaining why we have no problems with this form of the task. But in the first version of the task the subject matter is completely different and so is not dealt with by the cheater detector. Here we have to rely on poorly developed general reasoning capacities. In short, then, Tooby and Cosmides argue for the existence of specific modules concerned with belief fixation by appeal to the strengths and limitations of human reasoning. The logic of their argument is that the postulation of specific modules makes sense of patterns of strengths and weaknesses that would otherwise be mysterious.

The problem with this argument is that there is an alternative explanation of the experimental results. The reasoning tasks in Wason and Johnson-Laird's original experiments were of a highly abstract form. Now many people have difficulties dealing with abstract problems, finding it hard to understand what is being asked of them or what approach they should adopt. Up until recently, first-year undergraduates reading philosophy in most British universities were required to study formal logic. For many very bright students this proved to be a torment, and formal logic has gradually been replaced by more informal variants of logic that discuss reasoning in a more concrete setting. In my experience, students describe informal logic as more 'intuitive' than its abstract relation. Reflecting this, students who have difficulty in seeing that the argument 'if A then B, B, therefore A' is fallacious have no such problems when presented with an argument of that form that has a concrete subject matter (for example, 'if Angus lives in Glasgow then he lives in Scotland. Angus lives in Scotland. Therefore, Angus lives in Glasgow').

What this example of logic suggests is that many people find it easier to deal with problems that have a concrete subject matter than those that are presented in a purely abstract format. But note that, though they have a concrete subject matter, these examples have nothing to do with cheating or social exchanges. What this suggests is that the salient property of the cases discussed by Tooby and Cosmides that enables most people to deal with them effectively is not that they relate to cheating or social exchanges but that they have a concrete subject matter. But, if that is the case, then the motivation for postulating a cheater detection module evaporates.

A final argument for the massive modularity thesis is that a competent human is able to perform a wide range of tasks each of which places quite different demands upon them. For example, predicting the behaviour of an inanimate object is quite different from predicting the behaviour of another human, which is in turn quite different from doing arithmetic or relocating one's car after a shopping trip in an unfamiliar town centre. What explains our ability to perform such a wide range of tasks is, so the argument goes, the fact that we have a distinct module corresponding to each of them.

One problem with this argument is that there are examples of tasks that place substantial and specific demands on us which most people perform very well, despite the fact that they couldn't plausibly be supported by a dedicated module. For example, consider reading and writing. Most contemporary Westerners learn to read and write with ease considerably before they leave primary school. Yet, as reading and writing are relatively recent cultural inventions, it is not plausible that our hunter-gatherer ancestors evolved a module for reading and writing in the Pleistocene. But if our minds are such as to enable us to master certain complex tasks without the possession of a dedicated module, then perhaps they enable us to master many more of our cognitive capacities in the absence of a corresponding module.

As it stands, the domain-specificity argument doesn't support the massive modularity thesis. But perhaps the considerations to which it appeals gives some reasons to think that there are modules over and above Fodorian input and output systems. I cited reading and writing as capacities that don't plausibly have an associated module. But there are a couple of things to note about reading and writing. Not every cognitively normal individual develops such capacities, and those who do develop them slowly and on the basis of much specific instruction and practice. There are also many other skills about which similar points can be made: the ability to play chess, cook, and so on. All these abilities rely upon knowledge (knowing the rules of chess, the correlation of symbols to sounds, and so on). Such knowledge relates to a domain of reality independent of the individual, though that reality might relate to social conventions and rules. In short, some of us have knowledge about certain specific domains of reality; that knowledge is both learned and supports specific cognitive capacities and isn't associated with a distinct module.

5 Psychology and physics

Suppose that there are domains of reality about which all normal humans have knowledge irrespective of their social or cultural background or whether they have been recipients of directed learning and teaching. Also suppose that that knowledge was of fundamental use and value to us and that it was also present in other creatures. Finally, suppose that the relevant domains of reality were governed by quite distinctive principles. Then, from an evolutionary point of view, it would make sense for such knowledge to be innate or easily acquired by means of domain-specific learning mechanisms (as opposed to domain-general learning mechanisms) and to be associated with a specific module.

Some prominent work in developmental psychology suggests that there are such domains of reality about which children have specific knowledge very early in their lives, raising the possibility that we have associated modules. Spelke (1994, 2003) writes of our having core systems, and other developmental psychologists (for example, Carey, 2009) talk about core domains of knowledge or theories (Gopnik and Meltzoff, 1997). Frequently discussed domains relate to inanimate physical objects and minded agents (particularly, other humans). Consider each of these in turn.

Inanimate objects behave in certain regular, law-governed ways, and having knowledge of such regularities is crucial if we are to predict the behaviour of these objects and successfully manipulate them. Spelke (1994) believes that we have a core system for representing and engaging with inanimate physical objects. She describes this system in such a way as to suggest that it is a module in the relevant sense of the term: a domain-specific and informationally encapsulated subsystem of the mind. Central to this module is the representation of a number of principles concerning the behaviour of inanimate physical objects. These principles can be described in the following terms.

1 *The principle of cohesion*: objects are connected masses of stuff that move as a whole.
2 *The principle of solidity*: objects are not easily permeated by other objects.
3 *The principle of continuity*: objects move in continuous paths travelling through space without gaps.
4 *The principle of contact*: objects move through contact with other objects.

Experiments on children suggest that such knowledge is in place very early on, before the child has had any chance to learn it or the abstract concept of an object that is central to it. To get a flavour of such experimental work, consider the following experiment carried out by Karen Wynn (1992), described in Spelke (2003). Wynn conducted an experiment on five-month-old infants involving a stage, two puppets and a screen. Initially one puppet is placed on the stage in full view of the infants. Then a screen is lowered to hide the puppet from view before a second puppet is placed behind the screen in such a way that the infants can see this manoeuvre. Finally, the screen is lowered. On some occasions the screen is lowered to reveal two puppets and on others to reveal just one (in the latter scenario, one of the puppets is surreptitiously removed from the stage before the screen is lowered).

The point of the experiment is to determine which of these two possibilities most surprises or violates the expectations of the infants. If the infants are most surprised when only one puppet is present, then that suggests that they think of inanimate physical objects as existing over time and as continuing to exist when they are not being perceived. If they do not think of the puppets as objects that continue to exist when unperceived, then they will be most surprised to find two puppets on the stage, as this is an experience they have never had before (recall, at the beginning of the experiment one puppet was placed on the stage so the infants have experience of the one-puppet scenario).

This raises the question as to how we are to determine the degree of surprise of the infants, given that they are pre-linguistic and so cannot tell us of their expectations. Wynn used looking time to determine expectations on the assumption that an infant will look longer at a situation that surprises her than one that she expected. What Wynn discovered was that the infants looked longer when only one puppet was present. From this she concluded that the infants had an innate concept of a physical object as something that persists over time even when not being perceived and a knowledge that the world is populated by such objects. Such a concept and associated knowledge is regarded as innate as it is present at too early a stage in the infants' lives to have been learned on the basis of experience.

Inanimate physical objects are prominent members of our world but don't exhaust what it is worth knowing about. In addition, there are people who are minded and whose behaviour is goal driven and dependent on their state of mind so that they are capable of initiating their own motions. An individual who predicted the behaviour of other people in the same way as they did inanimate physical objects would make systematic errors (Bloom, 2004; Epley, 2014).

There is evidence that children draw a distinction between people and inanimate objects very early in their life and have different expectations concerning how they behave. For example, though infants express surprise when they see an inanimate object move without seeming to be subject to the contact of any other object external to it, they become disconcerted when a previously mobile human face falls still (Tronick et al., 1978). This suggests that they have a quite different core system for dealing with other people, one that utilizes different representations and principles from that implicated in their dealings with inanimate physical objects. For Spelke (2003) and others (for example, Boeckx, 2010), this core system is a module.[14]

But how plausible is the idea that we have modules for dealing with inanimate physical objects, on the one hand, and minded agents, on the other? Although I am sympathetic to the idea that we have an innate concept of an object and innate knowledge of principles such as those identified by Spelke,[15] I am sceptical of the idea that our dealings with inanimate physical objects are based upon a dedicated module. The first thing to note about Spelke's principles is that they are very general in nature. That is, they are supposed to apply to all inanimate physical objects regardless of their shape, size or constitution. No doubt, knowledge of such principles would give an infant a head start in her dealings with the external world, but they wouldn't allow her to predict the behaviour of inanimate physical objects with much precision or manipulate such objects effectively. For example, suppose I am holding an object and then I let go of it. What will happen next? It is easy to predict that the object will fall to the ground on the basis of knowledge of the general principle that unsupported objects fall. But will it drop suddenly or slowly flutter to the ground? And when it hits the ground will it shatter, splatter, bounce or just land with a thud? One way of making accurate predictions about such matters would involve utilizing knowledge of more specific principles that relate such properties of an object as its shape, size, mass, elasticity, constitution, and so on, to its behaviour. There is little evidence that infants have knowledge of such specific principles, and even if they did these principles would be of little help to them. This is because the employment of such principles in accurate prediction requires that one has knowledge of the relevant properties of the objects in question. Now infants can be expected to have knowledge of the shape and size of objects in their environment but are rarely in a position to know of their specific mass, elasticity, constitution, and so on.

This raises the question of how infants are able to make specific predictions about the behaviour of the inanimate physical objects in their environment. My suggestion is that, initially, they cannot make anything more than gross predictions, as would be expected if all they had to rely on were Spelke's general principles. But on the basis of experience they acquire a rich body of knowledge that eventually enables them routinely to generate precise and accurate predictions. The appeal to experience in acquiring knowledge doesn't in itself imply that there isn't a module for dealing with inanimate physical objects. After all, those who postulate a language module believe that it contains lexical information that is learned (information about the words that belong to the language and what they mean). But there would appear to be a difference between such lexical knowledge and knowledge of the behaviour of physical objects relating to its accessibility and relationship to other bodies of knowledge. The word 'dog' belongs to my language, but ask me what that word means and I will flap and flail, and any answer that I give will hardly do justice to the intricacies of that meaning as represented in my head. Thus it would seem that knowledge about meaning is largely unconscious and requires the hard work of the theoretical linguist to uncover it. What we do have more direct access to, on the other hand, is knowledge about dogs, the type of animal picked out by the word 'dog'. For example, I know that dogs are descended from wolves, that they tend to bark, wag their tail when happy, chase postmen, and so on.

Suppose an infant comes across a yellow-green spherical object of about 12 centimetres in diameter with furry texture. What will it do if dropped from chest height? In virtue of its shape, size and mass it could reliably be expected to fall to the ground, and an infant with little experience of the world armed with a knowledge of principles such as Spelke's may well have this expectation. But what will it do once it hits the ground? Nothing about the readily perceivable properties of the object tells us whether it will bounce, shatter, splatter or whatever. However, I am firmly confident that the object would bounce to roughly half the height from which it was dropped and would be very surprised if it shattered or splattered. I am confident of this because, on the basis of its appearance (and perhaps the context in which I see it), I categorize it as a tennis ball, and I know that that is generally how tennis balls behave and that objects that aren't tennis balls rarely look like tennis balls. Though if I came across the object in a joke shop I might not expect it to bounce, because I know that many of the items to be found in joke shops are made to violate our normal expectations. An infant can hardly be expected to know all this and will have to have considerable experience of tennis balls to make any reliable predictions about its behaviour.

My point generalizes. The kinds of physical properties that we typically perceive objects to have do not in themselves enable us to predict how they will behave in any precise respect. There is no law of physics that says that spherical, furry objects with a diameter of 12 centimetres will bounce. To be able to make specific and reliable predictions we have to be able to categorize things on the basis of their readily perceivable physical properties (and, perhaps, the context in which they are located) and have a good deal of learned information about how the items that belong to such categories typically behave. Thus, I have learned about the distinctive appearance of tennis balls, where they tend to be located and how they typically behave, and it is this, rather than any general knowledge of physics, that grounds my prediction that the object will bounce. And I have learned comparable information about glasses, potatoes, paper, marbles, and so on. In sum, then, the knowledge that enables us to predict how inanimate physical objects behave is not knowledge of principles about how physical objects in general behave but quite specific knowledge about the behaviour of particular types of objects, along with knowledge that enables us to identify things as belonging to those specific types.

A number of points can be made about this knowledge. First, much of it is learned; it is barely plausible to claim that we have innate knowledge about tennis balls (that such things exist, what they typically look like, where they are normally located and how they behave). Second, such knowledge, unlike that concerning the meaning of words, is readily accessible to consciousness. Third, such knowledge is acquired by means of domain-general rather than domain-specific learning mechanisms. For example, it is barely plausible to postulate a tennis or a sports module to explain how we acquire knowledge about tennis balls. Fourth, such knowledge can be used in a wide range of cognitive tasks and is readily combinable with other knowledge items. What these points about our knowledge of tennis balls – and a whole battery of types of physical objects for which we have concepts – imply is that such knowledge doesn't belong to any domain-specific module. Thus, the bulk of our specific predictions about the behaviour of inanimate physical objects isn't based upon the operations of a module of the kind that Spelke postulates.

Strictly speaking, none of this implies that the module that Spelke postulates doesn't exist. But it does imply that, if there is such a module, it plays only a limited role in our cognitive lives once we are beyond an infant's state of ignorance. Either it becomes dormant, or it is utilized to give gross predictions when dealing with unfamiliar objects. However, in the light of this, perhaps we should take seriously the suggestion that Spelke's principles don't belong to a domain-specific module but constitute a body of quite general innate knowledge that is supplemented by more specific knowledge acquired by means of learning. In this connection it is worth pointing out that other developmental psychologists who are quite happy to talk about core knowledge domains, and to credit us with some innate knowledge relating to those domains, do not think of them in modular terms. Rather, they think of the innate aspects of such knowledge as provisional theories that are subsequently overwritten by more sophisticated theories acquired by means of domain-general learning mechanisms. For example, Gopnik and Meltzoff (1997) portray children as 'little scientists' who use the very same cognitive mechanisms as are used by professional adult scientists to develop theories across a range of domains.[16] And Susan Carey (2009) argues that we use 'Quinean bootstrapping' to develop representational systems across a range of domains that go beyond and are incommensurable with innate core cognitive systems.

Human individuals are unlike inanimate physical objects in that they have minds and act on the basis of their mental states. The ability to work out the mental states of other people and predict and explain their mental states and actions on the basis of such attributions is generally known as the capacity to mind-read or the theory of mind (Epley, 2014). Is the capacity to mind-read based on a specific module? Do we have an encapsulated theory of mind module? In addition to Spelke (2003), a number of prominent philosophers and cognitive scientists think that we do.[17] However, for considerations parallel to those that I presented in relation to the postulation of a module for dealing with inanimate physical objects, I am sceptical.

One prominent debate about the capacity to mind-read is that between so-called theory theorists and simulation theorists. The former think that the capacity to mind-read is based upon a theory of human psychology that we utilize when mind-reading. The latter, on the other hand, think that when we attempt to work out the mental state of another person we do so by simulating them. To see what this amounts to, consider how we might go about predicting the behaviour of another person. According to radical simulationists such as Gordon (1986) and Heal (1986), predicting another person's behaviour can involve imaginatively projecting ourselves into their circumstances so as to find out how we would behave were we in the same position. Operating on the assumption that our mind works like that of everybody else, this enables us to reach a conclusion as to how the simulated person will behave. Alternatively, Goldman (1989, 2006) argues that we take our own decision-making procedure offline, feeding it pretend beliefs corresponding to those we take the person in question to have. We then conclude that the person will behave in line with the intention generated by our offline decision-making system in response to the pretend beliefs. Once again, this involves assuming that other people have minds that work just like our own.

I don't have space to do full justice to the simulationist approach and will keep my comments brief. I don't doubt that we sometimes engage in simulation activities when mind-reading, but I suspect that we do so in extreme circumstances when our normal resources break down (What would Bill do if he was made redundant? I've no idea. Well what would I do?) or when we are trying to morally evaluate someone's behaviour (Would I have done that in his circumstances?). In other words, simulation is a resource of last resort. One problem with simulation is that it can only be effective when dealing with people who are like us in the relevant respect. It's no good my resorting to simulation when trying to predict what you will order in a café if your tastes in coffee are very different to mine. Sometimes simulationists talk about making adjustments for different beliefs or personalities, but the question to which this gives rise is how do we determine the relevant differences of personality or belief if not by employing a theory? In sum, then, I am sceptical that simulationism can offer more than a partial account of our mind-reading capacities.

Simulationism is inconsistent with a modular account of mind-reading, as it rejects the idea that mind-reading is based upon the utilization of a body of domain-specific and encapsulated information. However, advocates of the theory theory view are not thereby committed to the existence of a theory of mind module. But how convincing is the modular version of the theory theory?

A first question to ask relates to the nature of the theory that is supposed to be encoded within the theory of mind module. Here there is a massive contrast with work relating to the language module. Cognitive scientists sympathetic to the idea of a language module[18] have produced a detailed account of its contents which goes a considerable distance towards explaining how it could play the role in cognition that is attributed to it. But the same could not be said with respect to the theory of mind module. What is generally claimed is that the module contains a theory made up of generalizations concerning the causal relations between various types of mental states and behaviour. Such generalizations quantify over the content of the mental states to which they advert. For example, one central generalization is this: if an individual wants p and believes that the best way to get p is to do q, then, all else equal, they will do q.

My worry is that, if a modular theory consists of such generalizations, it is difficult to see how it could routinely help us in our dealings with other people. A first problem is that for such a theory to be of any use we need to be able to feed it specific details concerning the propositional attitudes of the individual in question. Yet what we want a theory of mind module to do is help us to uncover the propositional attitudes of our fellows. Perhaps there are ways of getting round this problem. For example, what people say is a rich source of information about what they believe and want, so the theory could incorporate a generalization to the effect that if someone uttered a declarative sentence with the content p, then, all else equal, they believe that p. (There might also be generalizations that tell us how to detect sincerity.) At this point I'm not going to conclude that a theory of the kind that advocates of a theory of mind module have in mind couldn't ground our mind-reading capacities. But what I will say is that it is difficult to adjudicate the viability of their position in the absence of a detailed account of the form of that theory.

A second worry relates to my reflections on the postulation of a module relating to inanimate physical objects. There I suggested that our predictions are often based not so much on the employment of a general theory as on our learned knowledge of the typical behaviour of familiar types of object and our assignment of particular objects to those types on the basis of readily available perceptual information. For example, I predict that the yellow-green furry sphere will bounce when dropped, whereas I predict that the similar-sized orange sphere with a waxy pitted surface will splatter. This is not because I have access to a physical theory that tells me that differences in surface colour and texture are the causal basis of such differences in behaviour. Actually, my physical theory tells me that such properties are inert in this context. What is significant about the colour and the texture is that they help me identify the objects as, respectively, a tennis ball and an orange, so prompting me to draw upon specific knowledge of the behaviour of such objects – behaviour that has been acquired by means of general learning mechanisms and, rather than being encapsulated, is accessible to a wide array of different cognitive processes.

My suggestion is this. Just as we often do not draw upon an abstract and general physical theory in predicting the behaviour of physical objects, even if we have such a theory, we often do not (or do not need to) resort to an abstract and general psychological theory when mind-reading. To see this, consider the following example. I go into a bar and I see two groups of people, each four in number. One group is made up of men whom, on the basis of their appearance, I classify as a group of builders who have just completed a hard day's work. The other group is made up of women, whom I identify as being in their thirties, professionally successful and affluent. Now suppose a member from each group rises from their seat and heads towards the bar. What do they want and what will their subsequent behaviour be? I have no problems reaching the conclusion that both of the individuals are heading to the bar to buy a drink and that they will buy a drink for each member of their community. Why am I so confident of this? Well, I know that bars are places where people go to drink with their friends and that in the UK there is a tradition or convention of group members taking turns in buying drinks for everyone in their group. What will each person order? I would expect the man to order four pints of beer of some form. I would be very surprised if he ordered any wine (especially rosé wine). On the other hand, I would predict that the woman will order four glasses of wine (and, as it is the height of summer, at least some of those glasses will be either of rosé or white wine). I would be very surprised if any pints of beer were ordered. In short, I come to certain conclusions about what these people want and how they will behave in the near future. But, rather than employing a general abstract psychological theory to do this, it would seem that I draw upon knowledge of what generally goes on in certain specific settings, along with knowledge of the typical appearance, lifestyle and mind-set of individuals belonging to particular professional groupings. Such knowledge, of which I have lots, would appear to be learned and, rather than being contained in any encapsulated module, is part of my general stock of knowledge of the world accessible to a range of distinct cognitive processes. My point is that such knowledge enables us to go a long way in working out the mental states and predicting the behaviour of our fellows. That is not to say that we don't have a general psychological theory or that it is not encoded in a theory of mind module. But any such theory or module need not be central to our everyday mind-reading activities.

6 Conclusion

In this chapter I have examined various versions of the modularity thesis, chief among them being, first, Fodor's idea that the input systems involved in perception are modules and, second, the massive modularity thesis, according to which even higher-level cognition is grounded in the operation of modules, of which there are many making up the human mind. None of these positions have been either decisively vindicated or refuted, and so the debate is very much alive and ongoing. My sympathies lie with the view that input systems involved in perception are modules and that there is a module dedicated to language.[19] As for central cognition, perhaps we innately carve up the world into distinct domains, including the psychological and the physical, and have associated innate knowledge as to the operations of those domains. However, I doubt that such knowledge would enable the growing child to prosper were it not supplemented with a large battery of knowledge learned on the basis of experience – knowledge that was generally available for a wide range of cognitive tasks.

4 Concepts

1 Introduction

Concepts play an important role in our cognitive lives, as we employ concepts whenever we have a thought, engage in reasoning or categorize an object. Without concepts we wouldn't be fully fledged thinkers, and an individual's stock of concepts limits the thoughts that she is capable of thinking. In the light of this, it should come as no surprise that questions about concepts have been very prominent within cognitive science. In this chapter I will examine the most prominent attempts within cognitive science to uncover the nature of concepts and how they are represented in the mind-brain.

2 What are concepts? Some preliminaries

What are concepts? Before addressing the contemporary theories that attempt to provide a substantial answer to this question, it would be helpful to have a preliminary answer acceptable to all that would serve to pin down the topic of discussion. As concepts are involved in having a thought, reasoning and categorizing, a natural suggestion is that concepts are the units of thought. That is, concepts are elements or components of thoughts in a way analogous to that in which words are components of sentences. Here I am using the term 'thought' in the standard contemporary philosophical sense to refer to propositional attitudes – that is, mental states such as beliefs, desires, intentions, hopes, fears, expectations, and so on (Bayne, 2013). On this view, for example, just as the word 'coffee' is a component of the sentence 'coffee keeps one awake', so the concept COFFEE is a component of the belief that coffee keeps one awake. This would imply that concepts are mental entities, things that exist in the mind. The problem with such an analysis is that many philosophers would disagree. This is because there is a rich tradition associated with Frege ([1892] 1980) that portrays concepts as abstract objects rather than mental entities. There are various motivations for such a view,[1] but a chief one is to avoid the implications that concepts cannot be shared or that one cannot know what concept another person expresses by a given word (and thus what they mean by that word). Such reasons for resisting the view that concepts are mental entities are bound up with the idea that mental entities are private, either in the respect that if something resided in my mind then it couldn't reside in your mind or in the respect that we can never know what resides in the mind of another person. I think there are good reasons for rejecting such a Fregean view of the consequences of the idea that concepts are mental entities, but I won't go into these here as there is a more pressing concern. If some philosophers and cognitive scientists view concepts as mental entities whereas others view them as abstract objects that exist outside of the mind, then it might appear that we have a fundamental disagreement, the upshot of which is that different theorists express different concepts by means of the word 'concept'. This would be an unsettling conclusion but fortunately it can be avoided. Even though Fregeans typically hold that concepts don't exist in the mind, they do think that we bear psychological relations to concepts. For example, we grasp concepts and our doing so is a precondition for having thoughts. In grasping the concept COFFEE, for instance, I stand in a psychological relation to that concept, and without doing so I would not be able to have the belief that coffee keeps me awake. Thus, two distinct people can share a concept by standing in the same psychological relation to one and the same external abstract object. This enables us to say that it is accepted all round that concepts are involved in having thoughts, reasoning and categorizing objects, and to regard the dispute as to whether concepts are mental or abstract objects as not being such is to imply that the combatants in that debate are talking past one another.

What I have said so far hints at a close relationship between concepts and language, as I have said that we use words to express concepts. Although there are controversies concerning the relationship between language and concepts, the following is fairly uncontroversial. What enables us to use words to express concepts relates to meaning. Just as words are meaningful so are concepts, although philosophers tend to use the term 'content' when talking about the meaning of a concept. So, for example, the word 'aard-vark' on my lips has a particular meaning and the concept AARDVARK has a corresponding content. With respect to a general noun such as 'aardvark' it is natural to think that a core element of its meaning is what it refers to or its extension: the word refers to a certain kind of animal (namely aardvarks) and all and only members of that kind of animal fall within its extension. Similarly, the concept AARDVARK applies to, and only to, aardvarks; all and only such animals fall under the concept, and that this is the case is a core element of the content of the concept AARDVARK.[2]

Another point of contact between concepts and words is that, just as words can be combined to form larger meaningful entities (such as phrases and sentences), so can concepts. For example, just as several distinct words can be combined to form the phrase 'the ferocious aardvark' and the sentence 'the ferocious aardvark bit the naturalist', so several distinct concepts can be combined to form the complex concept THE FEROCIOUS AARDVARK and the thought THE FEROCIOUS AARDVARK BIT THE NATURALIST. And, arguably, just as the meaning of a complex linguistic structure depends on the meaning of its basic components and how they are put together, the content of a complex conceptual structure depends on the content of its basic constituents and the way that they are put together (Frege, [1914] 1979). It is the fact that concepts have content and that they can be combined into larger contentful units that enables them to play their role in having thoughts, reasoning and categorizing.

We now come to a point of disanalogy between words and concepts. Concepts are inherently contentful not just in the respect that for something to be a concept it must have content. In addition, the content of a concept makes it the concept that it is in the respect that had it had a different content then it would have been a different concept. In other words, concepts are (at least partly) individuated by their content. It doesn't seem that the same is true of words, as one and the same word can mean different things on the lips of distinct individuals or a word can change its meaning over time. Thus, although words have meaning they are not individuated in terms of their meaning. On this point there is quite a lot of imprecision in the literature on concepts that I think we ought to be wary of. Sometimes philosophers and cognitive scientists talk of a concept changing its content. For example, they might talk of a concept widely shared within a community as changing or developing its content in the light of scientific advance. Or they might talk of an individual as having a concept that changes or develops in terms of its content as she develops a richer knowledge base and understanding of the world.[3]

I'm going to dig my heels in and insist that concepts don't change or develop their content. If a particular concept changed its content, then it would become a different concept. This is not to say that there cannot be change or development with respect to concepts in relation either to a community of thinkers or to an individual. To see this, it would be helpful to employ Georges Rey's (1994) terminological distinction between concepts and conceptions and note that sometimes people use the word 'concepts' when what they are really talking about are not concepts but conceptions.

A conception relating to a particular concept is a theory or body of beliefs about the items falling under the concept. This could be a theory about the nature of the items falling under the concept in virtue of which they fall under the concept. For example, I believe that, for something to be a sample of water, to fall under the concept WATER, it is necessary and sufficient that it is composed of molecules of H_2O, and I regard this as capturing a fact about the nature of water. Alternatively, it could be a belief that the items falling under the concept had a particular feature where that feature was not regarded as essential to falling under the concept. For example, I think that philosophers are rarely good at sport, but I don't think that it is essential to being a philosopher (essential for falling under the concept PHILOSOPHER) that one is bad at sport.

With respect to many of an individual's concepts, the associated conceptions are rich. That is, they contain an extensive network of beliefs. But note that, once we make this distinction between concepts and conceptions, the possibility opens up that distinct people can associate different conceptions with one and the same concept. For example, it would be a difference between the respective conceptions that you and I associate with the concept AARDVARK if you believed that they were herbivores while I believed they were insectivores. This would explain how we could have a debate about aardvarks without talking past each other. It also opens up the possibility that a person or a community can change their conception associated with a given concept. Often this will involve theoretical enrichment, as when I find out more about aardvarks by reading natural history books and visiting zoos. Often when we talk about a person or community's changing or developing a particular concept (as in 'with the chemical revolution of the late seventeenth century our concept of WATER changed' or 'as she grows older a child's concept of HEAT undergoes considerable development') what we are really talking about is a change or development of a conception associated with a given concept where that concept remains fixed.

Another point to note is that conceptions can be thin or undeveloped. The concept PROTEIN belongs to my conceptual repertoire, but what I believe about proteins doesn't amount to very much and could be written on a small sheet of paper in normal sized handwriting. Moreover, conceptions can contain falsehoods. Aardvarks are termite eaters, but that fact doesn't rule out the possibility of someone believing that aardvarks are herbivores.

It might be objected that my drawing a distinction between concepts and conceptions betrays a problematic commitment to concept atomism. Concept atomism is the view that the identity of a concept doesn't depend on its relations to any other concepts. To understand this, consider the concept AARDVARK and what a non-atomistic view of that concept would amount to. According to the non-atomist, the concept AARDVARK belongs to a collection of concepts to which it bears a special relationship – a relationship that makes it the concept that it is. Probable candidates for other members of this collection include the concepts ANIMAL and QUADRUPED. The relationship that AARDVARK bears to these other concepts is such that a person couldn't have the concept AARDVARK without also having those of ANIMAL and QUADRUPED and without being disposed to infer from the belief that something was an aardvark the belief that it was an animal and a quadruped. For the atomist, in contrast, the concept AARDVARK bears no special relationship to any other concepts, so there are no other particular concepts that a person must have in order to have that concept.

Why might one be tempted to endorse an atomistic view of concepts? The most prominent view in favour of atomism was developed by Jerry Fodor (1987, 1990). He argued that there is no stable resting place between atomism and its polar opposite holism. Holism is the view that the identity of a concept is determined by the relationships it bears to every other concept in the conceptual scheme to which it belongs. This implies that, if two people appear to differ with respect to a belief involving a particular concept (for example, one accepts the belief as true whereas the other regards it as false), then in actual fact they don't share the concept. For example, if I have a belief that I express by saying 'aardvarks eat termites' and you disagree, then whatever concept you express by means of 'aardvark' it isn't the same as what I express by that word. Indeed, the consequences of this disagreement would ramify throughout our respective conceptual systems, with the upshot that we didn't share any concepts. Why is this? Well it would immediately imply that we express different concepts by means of the words 'eat' and 'termite'. And if you and I go on apparently to agree that 'termites are animals', given that we express different concepts by means of 'termite', the upshot will be that we will express different concepts by means of 'animal'. This will in turn lead to conceptual diversity with respect to the concepts expressed by the words belonging to any subsequent sentence containing 'animal' that either of us asserts, and so on. As no two individuals are ever likely to share all their beliefs, holism implies that no two individuals are ever likely to share a single concept. For Fodor this is a disaster; of course distinct individuals often share concepts, even though they differ in their beliefs involving those concepts, so holism must be false.[4]

At this point one might object that there is a stable position lying between atomism and holism that we might call molecularism. According to such a view, for example, the concept AARDVARK bears relations to ANIMAL that it doesn't bear to TERMITE so that, although one couldn't have the concept AARDVARK without believing that aardvarks are animals, one could have that concept while not believing that aardvarks eat termites. In other words, concepts, rather than existing as isolated units or living in massive networks, come in small interrelated clusters. Fodor's objection to such a view is that there is no principled way to distinguish between those relations that a concept bears to other concepts that determine its identity from those that don't.

I'm not going to attempt to settle this debate between atomism, molecularism and holism now, for my primary concern was the accusation that, in drawing the concept/conception distinction, I had committed myself to atomism. The idea behind this objection is that I permit two individuals to diverge considerably in the theories and beliefs they hold in association with a given concept while still sharing that concept. My response is that, though my line of thought might not be consistent with holism, it is consistent with molecularism. This is because conceptions can be expected to contain beliefs that carry potentially significant information about the items falling under the associated concept without specifying a conceptual truth. For example, my belief that aardvarks eat termites is both true and a significant piece of information about aardvarks, but it does not correspond to a conceptual truth or a partial definition of what it is to fall under the concept AARDVARK.

3 Traditional theories of concepts

In the previous section I gave a preliminary account of what concepts are, an account that would be found relatively uncontroversial by most cognitive scientists. Going beyond this account involves constructing a substantial theory of concepts, and it is here that controversy and disagreement emerge. The last thirty years have seen the emergence of a number of competing theories of concepts, all of which are very much alive today. These are the prototype theory, the exemplar theory, the theory theory, informational atomism, and the proxytype theory. In this section I will provide an account of these theories, but before doing so I will set the scene by describing two traditional theories of concepts that are not widely held today. The point of discussing these theories is that they are historically important and intuitively appealing, and their perceived weaknesses served to motivate later accounts.

For all their differences, what binds together the two traditional and five contemporary theories of concepts that I shall discuss is that they all involve a commitment to the idea that concepts are mental entities. More specifically, they portray concepts as being mental representations of one kind or another. This puts them in conflict with the Fregean view of concepts as abstract objects that, as we have seen, is popular in some philosophical circles. Nevertheless, there is a way for the Fregean to view these theories which avoids the implication that she is an inevitable opponent of them all. This involves viewing them as theories of what is going on in the mind of an individual when she grasps a particular concept in virtue of which she grasps that concept. For, recall, even though the Fregean regards concepts as abstract objects, she holds that individuals stand in psychological relations to those objects and that, if they didn't do so, they wouldn't be capable of thinking. In other words, each of the seven theories of concepts can be viewed as an account of the psychological relation that we bear to our concepts.

The first of the two traditional theories I shall discuss is the imagistic theory. This theory identifies concepts with mental images. So, for example, employing the concept DOG would involve having an image of a dog before one's mind's eye. The imagistic theory is often associated with the British empiricist philosophers John Locke ([1689] 1975) and David Hume ([1738] 1978), and there is much to be said in its favour. First, it does seem plausible on phenomenological grounds. For example, if I were to ask you to think of a dog, you may well experience a mental image of a dog, an experience akin to seeing a dog. Second, the imagistic theory readily suggests an account of concept acquisition – that is, an explanation of how we acquire our concepts. According to that explanation, when I first saw a dog, that animal caused me to have a visual image of it. That image was then copied by my mind and stored in long-term memory in a form that I recall whenever I think about dogs. Third, it suggests an explanation of the content of our concepts. Recall that our concepts have meaning or content and that this is central to their identity as the concepts that they are. For example, the concept DOG is about dogs rather than cats or animals in general. Thus, when you employ that concept in thought you are thinking about dogs rather than cats or animals in general. Similarly, when you employ that concept in categorizing something you are categorizing it as a dog rather than a cat or (merely) an animal. But this raises a question for any theory of concepts that identifies concepts with mental representations: why does the representation have the content that it has rather than some other content or no content at all? The imagistic theory suggests an answer: content is based on pictorial resemblance, so that my concept DOG has the content it has because it closely resembles dogs (rather than cats or animals in general).

Despite these strengths, the imagistic theory faces major problems, and the recognition of these led to a decline in its popularity and the search for an alternative theory. First, there are many concepts that are abstract in nature and so seem ill-suited to having an imagistic format. Perhaps a concept such as DOG or RED could take the form of an image, as dogs and the colour red have a characteristic appearance. But the same could hardly be said of concepts such as VIRTUE, JUSTICE, TRUTH or NOUN. For what does virtue, justice, truth or a noun look like? Second, concepts are often general in that they apply equally to many distinct things. For example, the concept DOG has a content such that it can be used to think about dogs in general rather than any particular dog and can be used correctly to categorize any particular dog no matter what its appearance. However, if I entertain an image of a dog it must be of a dog with a particular colour, hair type, tail length, and so on. Thus, it will resemble some dogs more than others. So the question arises as to why it has the general content *dog* (so that it serves as the concept DOG) rather than the more specific *short-haired, long-tailed brown dog* (so that it serves as the concept

SHORT-HAIRED, LONG-TAILED BROWN DOG)? In short, the imagistic theory makes it difficult to see how concepts can be as general as they often are.[5] Third, the issue of the meaning or content of images is not as straightforward as it first appears. This is because images don't inherently mean what they mean, so that a given image could be used to mean distinct things. As Wittgenstein (1953) pointed out, a picture of an old man climbing up a hill could be used as a picture of an old man sliding down a hill, as it resembles both scenarios equally. Hence, what is needed to fix the content of an image is something non-imagistic lying behind the image. Now what might that something be? An obvious suggestion[6] is that it is a mental state, such as an intention to mean a particular thing by the image or a particular interpretation of the image. But intending to mean something by an image or interpreting it in a particular way involves employing specific concepts. But those concepts can't have an imagistic format, otherwise they wouldn't have the determinate content they need to fix the content of the image.

The second traditional theory of concepts I will discuss is the so-called classical theory. The classical theory is prompted by the idea that, when someone possesses a particular concept, they know the necessary and sufficient conditions for falling under that concept. Thus, concepts are complex representations that take the form of a definition. Consider a familiar example. The necessary and sufficient conditions for being a bachelor are being an unmarried adult male. Consequently, having the concept BACHELOR is a matter of representing the fact that, for any x, x falls under the concept BACHELOR if and only if x is an unmarried adult male. If there is no concept in an individual's conceptual repertoire that they define in this way, then BACHELOR doesn't belong to that repertoire.

The main problem with the classical theory is that few concepts can be analysed or defined in terms of necessary and sufficient conditions. That is why I shifted examples from DOG to BACHELOR: while I grasp the concept DOG, I have no idea what the necessary and sufficient conditions for being a dog are, and the same point holds for most of my concepts. A number of otherwise very different philosophers have made this point. Wittgenstein (1953) focused on the concept of a GAME, arguing that it is not possible to define what it is to be a game as games have nothing in common in virtue of which they are all games. Some involve competition, others do not; some are played by teams, others are not; some involve a ball, others do not; some are physically demanding, others are not; and so on. Wittgenstein concludes that GAME is a family resemblance concept: what binds together distinct games is that they instantiate a network of overlapping and criss-crossing similarities. A similar objection is made by Jerry Fodor (1981a, 1998), who claims that few lexical concepts[7] are definable or that, in other words, few lexical concepts have necessary and sufficient conditions for their application. In *Concepts*, Fodor (1998) argues that, despite the exertion of considerable energy by philosophers and linguists, there are few lexical concepts that have been given plausible definitions. Typically, whenever a philosopher or a linguist produces a definition of a target concept, one of her colleagues presents a compelling counter-example. From the repeated failure of attempts to produce such definitions, Fodor induces the conclusion that most lexical concepts are indefinable.[8]

4 The prototype and exemplar theories

Wittgenstein's reflections on the concept GAME impressed the psychologist Eleanor Rosch, who, in the 1970s, played a key role in the development of a hugely influential and still very current theory of concepts. This is the view that concepts are prototypes.[9] A prototype is a complex mental representation which, rather than specifying necessary and sufficient conditions, specifies the characteristics that any item falling under it is likely to have. For example, on this view, the concept DOG is a complex representation that specifies properties that dogs generally have, properties that something is likely to have if it is a dog. Examples of such properties might be those of having four legs, having fur, having a tendency to bark, and so on. Thus, the DOG prototype implicitly provides a description of a prototypical or stereotypical dog, and grasping the concept DOG is a matter of having this description encoded in one's mind. A prototype also includes a similarity metric so that, for example, determining whether an item x falls within the extension of DOG involves employing a similarity metric in order to determine whether x resembles the prototypical dog to a sufficient extent. An example of a very simple similarity metric would be a rule to the effect that, if something had a particular proportion of the properties that figure in the prototype, it falls under the concept in question. A Labrador or a Golden Retriever would be a serious candidate for a prototypical dog but, presumably, a Great Dane or a Pekingese would not be. That an individual would categorize a Pekingese, but not a Siamese cat, as a DOG reflects the fact that the employment of the similarity metric generates the result that the former, but not the latter, is sufficiently similar to the prototypical dog – that is, has enough of the properties specified by the prototype – to fall under the concept DOG. As the prototype theory emphasizes the importance of representing properties that are typically, but not necessarily, possessed by items falling under the target concept, it is often described as constituting a statistical or probabilistic approach to concepts.

A characteristic often attributed to prototypes by advocates of this theory relates to the weighting of properties that feature in a given prototype. To assign a property a particular weight is to identify its significance when working out whether something falls under the concept in question, and in this respect not all properties are equal. Intuitively, the greater the weight of a particular property, the more representing something as having that property pushes one towards categorizing it as falling under the concept in question. For example, most birds have feathers and few things that aren't birds have feathers. And most birds fly, but so do lots of other creatures (for example, bats, butterflies, bees and some species of beetle). If you knew this, then if the only information that you had about something was that it was feathered, you would be more likely to categorize it as a bird than if the only information you had about it was that it flies. Reflecting this, it would be natural to attribute the property of being feathered a greater weight than that of flying in relation to the BIRD prototype. Note that the weight of a property might not go hand in hand with its commonality in the population of items falling under the concept. For example, all birds are animals but not all birds fly. Yet, given that there are far more animals that aren't birds than there are fliers that aren't birds, knowing that something flies is better evidence for its being a bird than knowing that it is an animal. Thus, one might expect the property of flying to have a heavier weight than that of being an animal. The weighting of properties within a prototype will be reflected in its associated similarity metric. Thus, for example, calculating whether something is sufficiently close to the prototypical dog to fall under the concept DOG will involve taking into account the various weights of the properties that figure in the prototype that the thing is represented as having.

Prototypes are complex representations in that the prototype constituting the concept DOG will be made up of representations of properties that are basic relative to that prototype. It is possible for some of these basic representations to have their own prototypes. For example, the prototype for DOG might include a representation of the property of being furry that is basic relative to that prototype. But, in addition, there might be a separate prototype constituting the concept FURRY, made up of the representations of properties that are basic relative to that prototype. Some of these representations might in turn have their own prototypes. Thus, a body of prototypes could constitute a network of related representations. However, not all of the properties represented in the prototypes in an individual's mind can have their own dedicated prototype, as that would give rise to an infinite regress of prototypes. Ultimately, an individual's mind must draw upon a stock of primitive representations out of which it builds prototypes and for which it has no corresponding prototypes.

This general account of the prototype theory suggests it is a general framework that provides its sympathizers with considerable latitude, so we can expect to find much disagreement between them. For example, two advocates of the prototype theory might disagree with respect to any of the following issues:

- What are the primitive representations out of which our prototypes are ultimately constructed?
- For any given concept that is constituted by a prototype, what properties are represented by its components and what weights do they have?
- With respect to any given prototype, what form does the associated similarity metric take?

Another potential source of disagreement relates to the debate between classical computationalism and connectionism, as a prototype could be encoded by means of either a classical computer or a connectionist network. As I see it, much work within the prototype framework is neutral on this question.

The prototype theory has proved to be very popular for a number of reasons which go beyond its merely providing an alternative to problematic traditional theories. The primary focus of those cognitive scientists who developed the prototype theory was with categorization, and they were impressed by two phenomena that emerged in experimental settings (Murphy, 2002). The first relates to typicality judgements. It is commonplace for individuals to judge that certain items that fall under a given concept are more typical items than others. For example, when asked, most Westerners say that a robin is a more typical bird than a penguin, where making that judgement doesn't involve viewing penguins as borderline birds or any less of a bird than robins. Second, we are quicker at categorizing some items that fall under a given concept than we are at categorizing others. For example, Westerners typically categorize a robin as a bird much more quickly than they do a chicken.

These phenomena are something that cry out for explanation, and the prototype theory offers a ready explanation. Items that are judged as being typical are those that come close to the prototypical instance of the concept. In other words, they have a high proportion of the most heavily weighted properties that figure in the prototype. Moreover, it is to be expected that, the closer an item is to the prototypical instance of a concept, the quicker it will be categorized as falling under that concept. For example, if classifying something as a BIRD involves determining that it meets a certain threshold with respect to properties that figure in the BIRD prototype, then it will normally take less time to determine that a bird with a high proportion of the most heavily weighted properties is a BIRD than it will a bird with a lower proportion of such properties.

Another strength of the prototype theory relates to concept acquisition. Most philosophers and cognitive scientists think that, even if some of our concepts are innate, the vast majority of them are learned on the basis of our experiences. Thus, for example, you acquired the concept DOG on the basis of experiences of dogs, and someone who didn't have such experiences would find it difficult to acquire that concept. Experiencing a dog involves perceiving it through one's sensory modalities (be it vision, audition, olfaction, touch or whatever). Now one could perceive a dog without conceptualizing it as a dog or, indeed, without having the concept DOG. But perceiving a dog would involve representing it as having certain properties, properties to do with its appearance (its shape, colour, movement, sound, smell, and such like). According to the prototype theory, learning a concept involves building a relevant prototype on the basis of one's experiences. Now suppose one had a series of experiences of a range of different dogs over a period of time, each of which involved representing an individual dog in terms of its readily perceivable properties. You would then have had precisely the kind of experiences on the basis of which one could build a prototype that would be satisfied by dogs, and only by dogs.

Before looking at criticisms of the prototype theory, it will be helpful to examine a closely related theory, namely the exemplar theory.[10] Suppose that before a visit to the zoo you had never come across an aardvark and didn't possess the concept AARDVARK. However, during your visit to the zoo you encountered your first aardvark and were told that that creature was an 'aardvark'. Later on in the day you see a strange creature and ask yourself what it is. You then remember your earlier encounter and, judging that the creature before you is quite similar to the one you perceived earlier, you conclude that it is the same kind of creature, namely, one that is called an 'aardvark'. Now consider a different scenario. I ask you to think of a dog on the assumption that you possess the concept DOG. Many people, when asked to perform such a task, report that they think of a particular dog of their acquaintance – for example, the dog they had as a child, the dog they currently own, or whatever. Moreover, this thinking involves having an experience akin to one of perceiving the individual dog in question. In other words, in thinking about dogs one has an offline perceptual experience of a particular dog with which one is acquainted.

Phenomena such as those described in the previous paragraph motivate the exemplar theory of concepts. An exemplar is a perceptual representation of a particular thing one has experienced. Thus, an individual's concept DOG, rather than being a representation of a list of weighted properties, is a perceptual representation of a particular dog of their acquaintance. Or, alternatively, it is a collection of perceptual representations each of a distinctive dog of their acquaintance. Described in these terms, the exemplar theory might appear to be quite different to the prototype theory. However, there are a number of points of contact. First, the exemplar theory, just like the prototype theory, emphasizes categorization and portrays categorizing something as comparing it with a stored representation and determining whether it is sufficiently similar to the referent of that representation. Second, the exemplar theory explains concept acquisition in terms of perceptual experiences of items falling under the target concept. Third, the exemplar theory is subject to the same prominent objections that have been directed at the prototype theory. These objections serve to motivate alternative theories, and it is to them that I now turn.

A prominent and widely discussed objection to prototype theories has been developed by Jerry Fodor (Fodor, 1998; Fodor and Lepore, 1996). Fodor argues that it is a fundamental property of concepts that they compose – that is, that concepts can be combined to form more complex concepts where the content of the complex is determined by the content of its constituent parts and the manner in which they are combined. For example, the concepts BLACK and DOG can be combined to form the concept BLACK DOG, where the content of the latter is inherited from the contents of the former. However, argues Fodor, prototypes do not generally compose, as the prototypes of complex concepts often bear little relation to the prototypes of their components. Consequently, concepts have a salient property that prototypes do not have, and this implies that concepts cannot be prototypes. Fodor provides a nice example to support his argument. A goldfish is neither a prototypical pet nor a prototypical fish. A dog would be an example of a prototypical pet and a trout a prototypical fish. Nevertheless, a goldfish is a prototypical pet fish. Thus, the prototype for the complex PET FISH will contain representations of properties – such as living in a bowl, being less than 10 centimetres long, being gold in colour – that do not figure in the prototypes for PET and FISH.

Highly technical attempts by advocates of the prototype theory to show that prototypes do in fact compose are not uncommon, but Fodor argues that such attempts have no hope of working. This is because what counts as a prototypical instance of a complex concept depends upon contingent facts about the world that are independent of what count as prototypical instances of its component concepts.

How convincing is Fodor's argument? One interesting objection developed by Edouard Machery (2009) comes as part of a general examination of work on concepts in both philosophy and psychology. Machery argues that psychological and philosophical work on concepts have quite different explanatory ambitions and so cannot be evaluated by the same criteria. Psychologists are concerned primarily with the mechanisms involved in categorization, concept acquisition and inference (particularly inductive inference). Philosophers, on the other hand, focus on how it is possible for us to have thoughts – that is to say, propositional attitudes such as beliefs and desires. A core element of this project involves explaining how our thoughts manage to be about what they are about. Fodor would be a clear-cut example of someone whose work on concepts addresses a philosophical agenda, and the claim that concepts compose and the demand for an explanation of compositionality is central to that agenda.

But the prototype theory is constructed as part of a quite different psychological agenda of explaining categorization and concept acquisition. Consequently, it doesn't count against the prototype theory that it doesn't meet Fodor's demands. In effect, Machery is saying that to criticize the prototype theory as Fodor does would be like criticizing Dostoevsky's fiction for not containing enough gags.

I'm not convinced by this objection. For it to go through it would have to be the case that psychologists and philosophers were talking about quite different things when they used the term 'concept'. Indeed, Machery seems to be suggesting that this is the case, as he says that 'concepts in psychology' are 'bodies of knowledge that are used by default in the processes underlying the higher cognitive capacities' (2009: 7), whereas 'concepts in philosophy' are 'capacities for having propositional attitudes' (ibid.: 31). I don't deny that there are differences in the aims, emphases and methods employed by, respectively, psychologists and philosophers, yet Machery overstates the extent and significance of these differences. Historically philosophers interested in concepts have been concerned with how we acquire concepts, how we use them to categorize and how we make inferences involving them. The British empiricist philosophers Locke and Hume stand out in this regard. Moreover, it is difficult to see how psychologists couldn't be concerned with our capacity for thought. For isn't categorizing something as a dog a matter of thinking or believing that it is a dog? And isn't inducing from one's experience of several dogs barking that all dogs bark a matter of forming one belief on the basis of another? Of course a psychological theory of concepts doesn't have to explain every property of concepts. But a given theory is problematic if it implies that concepts don't or couldn't have a property which we have independent reason to believe that they have. And it is this thought that lies at the heart of Fodor's objection.

Nevertheless, I'm not convinced by Fodor's compositionality objection for the following reason. When Fodor makes the claim that concepts compose, he is talking about content; the content of a complex concept is inherited from the content of its components. The reason why he is insistent that concepts compose in this sense is because such compositionality provides a ready explanation of two prominent features of thought which we saw him emphasize in chapter 2 – namely, the systematicity and the productivity of thought. But, as it stands, he hasn't established that prototypes don't compose in the relevant sense. For all he has shown is that the features present in the prototype of a complex concept don't always correspond to those present in the prototypes of its constituents. But, for this to imply a lack of compositionality with respect to content, there has to be a direct relationship between the features present in a prototype and its content. However, it is far from clear that the advocate of the prototype theory is committed to the existence of such a direct relationship. To see this, consider the following. The features present in a prototype are themselves representations. What explains their content? One could coherently answer this question in Fodorian terms by saying that it is a matter of what reliably causes their activation or tokening. With respect to the prototypes in which such features figure, one could also argue that their content was a matter of what reliably caused their tokening or activation. Thus, if the feature BARKS figures in the DOG prototype, then the content of the former doesn't enter into the content of the latter. The upshot of this is that, if the prototype for a complex concept such as PET FISH contains quite different features than its constituent concepts PET and FISH, it doesn't follow that the content of the former breaks contact with that of the latter. But, one might object on Fodor's behalf, in such a case we can't identify any contribution that the concepts PET and FISH make to the content of PET FISH, as would be the case on Fodor's model or if the features of the complex concept were derived from the features of the constituents. But, the objection continues, this implies that the complex PET FISH is no more a complex concept than the concepts PET and FISH, and we are tempted to think of it as a complex concept only because the English word that expresses the concept is itself composed of the English words 'pet' and 'fish'.

In the light of this objection, what is needed is an alternative account of compositionality, and here is a sketch of such an account. We can distinguish between two ways of constructing a prototype. The first is on the basis of interactions with the environment and doesn't involve drawing upon other prototypes.[11] Hence, one adds features to the prototype under construction on the basis of determining the properties of items falling under the target concept. The content of such prototypes is a matter of what reliably activates them. A second way of building a prototype involves constructing one from pre-existing prototypes. So, for example, I might construct the prototype ANGRY DOG not on the basis of interactions with angry dogs but by means of a process that draws upon my prototype for ANGRY and that for DOG. In which case the content of the prototype depends on the content of those from which it was derived. In other words, the content of the prototype is a matter of its history. This suggestion might be seen to be unappealing because it introduces unnecessary complexity into the explanation of the basis of content or makes an ad hoc distinction. In response I would say that the complexity is no greater than in Fodor's account. For Fodor, FISH and PET get their contents directly from what reliably causes their tokening. PET FISH, on the other hand, doesn't have the content that it has because its tokening is reliably caused by pet fish but because it has one component whose tokening is reliably caused by pets and another whose tokening is reliably caused by fish. In effect, what I am suggesting is that what makes a prototype an expression of a complex concept is its historical origins, its being constructed from two distinct prototypes, both of which contribute to its content. For this to be the case there have to be mechanisms that combine prototypes. But there is no guarantee that the mechanism merely merges the features present in each of the original prototypes. As Prinz (2002) suggests in attempting to deal with Fodor's objection, such construction mechanisms might sometimes draw upon background knowledge, so influencing what features figure in the resultant prototype.

Another possibility arises concerning those cases where two prototypes conflict with respect to a particular feature. Suppose the COW prototype contains the feature DOCILE and that this is heavily weighted. This feature will be inconsistent with features present in the prototype for ANGRY, generating potential problems when constructing the prototype for the concept ANGRY COW. Consequently, when combining these two prototypes, the potential inconsistency must be detected and resolved by deleting the feature DOCILE or by reducing its weight to a very low level. In short, what I am suggesting is the possibility of mechanisms for constructing prototypes by combining prototypes that already exist in the mind. These do not serve to express complex concepts because they have distinct parts each corresponding to a distinct concept. Rather, they express complex concepts because of their historical basis in distinct prototypes, each of which was independently constructed on the basis of interactions with items falling under it.

I now come to another important criticism of the prototype theory (and, by extension, the exemplar theory). This criticism is bound up with, and motivates, an alternative theory, namely the theory theory of concepts. Consider the BIRD prototype. A prototype theorist would regard this as containing representations of the properties of (being capable of) flight and being winged. Thus, the prototype would represent birds as generally having wings and generally being capable of flight. But it wouldn't represent the relationship between these properties; in particular it wouldn't represent the fact that birds are able to fly *because* they have wings.

But, it might be objected, the upshot of this is that the prototype theory overlooks the important role of our concepts with respect to explanation. One of the things we use concepts to do is to explain why things are the way they are or why things behave the way they do. For example, we explain the ability of birds to fly in terms of their having wings. Often this explanation takes a causal form; having wings is the causal basis of a bird's ability to fly; the bird flew because it flapped its wings. Having the ability to construct explanations is a hugely important cognitive capacity, as it provides us with information that enables us to predict events and so plan how best to act to satisfy our needs and desires.

5 The theory theory of concepts

The theory theory places explanation at the heart of its account of concepts, and it does this through drawing an explicit comparison between concepts and scientific theories. In short, it says that concepts are theories that are akin to scientific theories.[12] To unpack this, it will be helpful to recapitulate some of the key features of science and scientific theories as described in chapter 1.

Here are some prominent features of science. First, most scientists assume that happenings in the natural world are not entirely random and irregular but are governed by laws. Thus, one of the main goals of science is to discover the laws that govern the workings of the natural world. Second, it is a goal of science to explain features of the natural world rather than merely to describe them, and laws are often appealed to in such explanations. Third, science is an empirical discipline in that observation and experiment play a central role in the scientific project. Fourth, in the course of explanation, scientists often postulate theoretical entities. These are things that are not observed but are invoked to explain phenomena that are observable. Prominent examples include genes, microbes, atoms, quarks and photons. Fifth, there is not just one science but many distinct sciences, each of which has its own distinct explanatory goals, theoretical commitments, research methods and technical vocabularies. Examples of distinct sciences are physics, chemistry, biology, meteorology, geology, psychology, and so on.

Given that science has the features described in the previous paragraph, we can give a general account of a scientific theory in the following terms. A scientific theory is a complex structure that purports to explain phenomena in some specific domain of the natural world. It will typically postulate unobservable phenomena that have the power to cause observable events and laws governing the interactions between phenomena (observable and unobservable).

The theory theory of concepts has its origins in developmental psychology, and so its champions are concerned primarily with the nature of children's concepts and how those concepts develop as an individual grows. Accordingly, children are portrayed as being little scientists who construct theories on the basis of their experience and whose theories change and develop over time in a manner that echoes the history of science. Although experience plays a key role in the development of a child's theories, advocates of the theory theory usually credit children with a substantial innate endowment with respect to theories. First, children innately carve up the world into distinct domains, including the biological, the psychological and the physical, and construct distinct theories to deal with each of these domains. Second, the cognitive mechanisms that children employ to construct theories are innate and, according to some, are the very mechanisms that adult scientists employ.[13] Third, children have innate starting theories, though those theories are often replaced in the course of development (Carey, 2009).

Some advocates of the theory theory have attributed a further innate dimension to the theoretical perspective of children.[14] For they have claimed that children are essentialists about many of the categories for which they have concepts. This is to say that children think that the items that belong to a particular category are bound together by having a common essence. An essence is a collection of properties which something must have in order to belong to the category in question and which are the underlying hidden causes of the readily perceivable properties of the category members. Thus, if a child were an essentialist with respect to the category corresponding to the concept DOG, she would think that anything falling under that concept did so in virtue of having the relevant hidden properties – properties that are causally responsible for surface properties relating to its appearance and behaviour.

This view that children are committed to essentialism is known as psychological essentialism, and there is considerable empirical evidence in its favour. To get a flavour of this evidence, consider Frank Keil's (1989) classic experiment. Keil showed children and adults a picture of a raccoon. When asked, these subjects answered that the picture was of a raccoon. They were then told that the pictured animal underwent a series of changes, including changes to its appearance (through dyeing its fur and plastic surgery), the insertion of a smell sac, and modifications to its behaviour. They were then presented with a picture of an animal resembling a skunk and told that it was of the original animal post-modification. When asked about the identity of the animal at this stage, children over the age of seven and adults systematically answered that, despite its appearance, it was a raccoon, thereby indicating that, for them, something's being a raccoon is a matter of its origins and/or hidden nature rather than its observable properties.

Typically, psychological essentialists regard children as holding a placeholder conception of essence – that is, children do not usually have any substantial views as to the precise nature of the categories towards which they adopt an essentialist attitude (Medin and Ortony, 1989). Psychological essentialism puts further pressure on the prototype theory of concepts for the following reason. The properties that tend to be represented in a prototype are observable properties, and if concepts were prototypes then falling under a particular concept would be a matter of satisfying the relevant prototype. Thus, for example, falling under the concept DOG would be a matter of satisfying the DOG prototype. But if children are essentialists then they will think that whether or not something is a DOG is a matter of its hidden properties, so that it is possible for something to be a DOG while not satisfying the DOG prototype or to fail to be a DOG while satisfying the DOG prototype. In short, the prototype theory is not consistent with how children view their concepts.

Now let us examine in a little more detail what is involved in the theory theorist's central claim that concepts are theories. As we have seen, a core idea is that from childhood we carve up the world into distinct domains and construct theories to explain phenomena in each of those domains. Plausible candidates for such domains include the biological, the physical and the psychological. Let's focus on the last of these domains. Being able to explain the behaviour of psychological entities, namely other people, requires having a theory of psychology. This theory will be a complex structure that postulates a network of psychological states which cannot be directly observed in another person, along with a number of causal principles specifying the relations of these psychological states to one another, to observable behaviour and to the external environment. Armed with such a theory, an individual will be able to reach conclusions about the psychological states of her fellows on the basis of observing their behaviour and their environment. This will in turn enable her to explain and predict their behaviour and psychological states.

A simple example of a causal principle lying at the heart of such a psychological theory might be that, if a person desires that p and believes that the best way to get p is to execute action q, then, all else equal, they will execute action q. Now, on the face of it, endorsing this principle involves employing the concepts of BELIEF, DESIRE and ACTION, in the sense that you can't hold a theory containing this principle if you don't possess those concepts. This might suggest that concepts are more basic than the principles in which they figure. However, for the theory theorist, this is not the case, as the principles implicitly define the concepts in such a way that one couldn't possess the concepts without endorsing the principles. In other words, having the concept BELIEF is a matter of endorsing principles concerning the causal relations between beliefs, other mental states and actions. The same holds for the concepts of ACTION, DESIRE and numerous other concepts for mental states, such as INTENTION, HOPE, FEAR, EXPECTATION, and so on. The upshot of this is that one can't possess the concept of one type of mental state without possessing the concepts of a whole battery of other mental states. In other words, psychological concepts come in clumps.

One objection that might be directed at the theory theory is that it has the implication that people rarely, if ever, share a concept. This objection derives from Jerry Fodor's (1987, 1990) criticism of causal role semantics, the view that a meaningful item (be it a word or a concept) derives its meaning or content from its causal relations to other meaningful items. Suppose that the theory you hold about beliefs differs slightly from the one that I hold about beliefs. For example, you hold that beliefs bear slightly different relations to desires and actions than I do. Then we won't share the concept BELIEF; if you have that concept, then I don't, and vice versa. Thus, if we both use the word 'belief' we will use that word to express different concepts; it will mean one thing on my lips and something else on your lips. But this difference relating to the concept BELIEF will ramify throughout all our psychological concepts so that we won't share any such concepts. A similar point will hold for all the other domains about which we have theories: any difference between the theories held by two distinct individuals will imply that they don't share any concepts relating to that domain.

Fodor regards it as non-negotiable that we share many concepts, and Prinz (2002) states that it is a condition for an adequate theory of concepts that it explains how concepts can be widely shared (this is the publicity constraint). Hence, for Prinz and Fodor, having the implication that I have described would be a major failing of the theory theory. There are three possible responses open to the theory theorist. First, she could argue that we do not differ in the theories that we hold. The problem with this is that it is rather implausible unless all the theories we hold are part of our innate endowment or learned by means of explicit instruction and so acquired by means of a mechanism that prevented divergence from one person to the next. But, although many theory theorists argue that we have innate starting theories, they also hold that such theories are quickly superseded by new theories and that the mechanisms that mediate such development find little role for explicit instruction. A second option would be to bite the bullet and accept that distinct individuals rarely share concepts. I follow Fodor and Prinz in finding such an option unpalatable. A third option would be to argue that distinct theories can be more or less similar to one another so that two individuals could share similar concepts. The challenge facing anyone who wants to endorse such a position is to develop a viable account of theory similarity.

6 Informational atomism and the proxytype theory

The objection that the theory theory implies that distinct individuals rarely share concepts brings me to the next prominent theory of concepts. This is informational atomism as developed by Jerry Fodor (1987, 1990). As we saw in chapter 2, Fodor is committed to the existence of a language of thought (LOT). Although LOT is not a public language such as English, Italian or Japanese, it shares key features of such languages. In particular, it has a battery of meaningful primitive symbols and syntactic rules for combining those symbols to form complex structures such as phrases and sentences. And the meaning of any such complex is determined by the meaning of its primitive components and the way they are put together (that is, the syntactic structure of the complex). Symbols can be realized in the brain. That is, just as a symbol of English can be physically embodied by means of a sound or a mark, a symbol of LOT can be physically embodied by means of a state of the brain. LOT is the vehicle of thought in that, whenever she tokens a belief, desire or any other propositional attitude, an individual will token a physically embodied sentence of LOT in her brain that has the appropriate content. For Fodor, concepts are symbols of LOT. To have the concept DOG, then, is to have a symbol in one's LOT that has the content *dog*. This raises the question of the basis of the content of LOT symbols: why does the LOT analogue of 'dog' have the content *dog* rather than some other content or no content at all? It is Fodor's answer to this question that makes his theory a version of informational atomism. To a first approximation, he thinks that the content of a LOT symbol is a matter of what reliably causes it to be tokened. So, for example, the LOT symbol DOG has the content *dog* because its tokenings are caused by dogs and only dogs. Or, more precisely, because it is a law that dogs cause the tokening of DOG. Fodor recognizes that, as it stands, this won't do, as tokenings of DOG are often caused by things that aren't dogs, as when one mistakes a fox on a dark night for a dog or one thinks about dogs as a result of thinking about cats. So, one might ask, why doesn't DOG have the content DOG-OR-FOX-ON-A-DARK-NIGHT or DOG-OR-THOUGHT-ABOUT-A-CAT?[15] Fodor's (1987, 1990) answer is that the dog–DOG causal relation is more basic than the other causal relations into which DOG enters, in that the latter depend asymmetrically on the former. That is, were it not the case that dogs caused tokenings of DOG, then it wouldn't be the case that foxes on a dark night (or thoughts about cats) caused tokenings of DOG, but not vice versa.

This theory is atomistic in that it rejects the thesis that the content of a concept is determined by its relations to other concepts, so that, at least in principle, one could have the content DOG without having the concept CAT, ANIMAL or any other particular concept. Thus, for Fodor, concepts are certainly not theories. However, it is important to note that Fodor is happy to allow that complex mental structures such as beliefs and theories (encoded by means of LOT sentences) mediate the content determining causal relations between concepts and what they represent. It is just that the content of those beliefs and theories doesn't enter into the content of the concepts in question. This explains why you and I could have quite different theories or beliefs about dogs yet still share the concept DOG.

Before considering any criticisms of Fodor's approach, it will be helpful to examine the final theory of concepts in my survey. This is the proxytype theory as developed by Jesse Prinz, drawing upon the work of the psychologist Lawrence Barsalou (1999). Prinz's stated aim is to develop a contemporary empiricist theory of concepts, and he draws a distinction between long-term and working memory. Thoughts are occurrent states as opposed to states that exist in the mind for lengthy periods of time. Thus thoughts reside in working memory. And as having a thought involves deploying a concept, then concepts also exist in working memory. However, there is a close relationship between working and long-term memory in that items occurring in the former are often constructed from resources stored in the latter. Indeed, such a relationship exists in the case of concepts. With respect to concepts, what exists in long-term memory are complex networks of representations. What binds together the elements of these networks are causal connections. The elements are causally connected in that activation of any one element of the network (an activation that involves its tokening in working memory) will typically cause the activation of some other element.

These networks stored in long-term memory correspond to categories of things in the outside world. For example, there is a network corresponding to dogs. Such a network was constructed over time on the basis of perceptual interactions with dogs. Moreover, the network is constructed out of representational primitives that are utilized by our various senses and so represent the kind of properties that we perceive objects as having. For example, these primitives have contents such as *red*, *edge*, *round*, and so on, where their content is a matter of what they causally co-vary with. Given that their basic representational elements come from a variety of sensory systems, the networks are multi-modal representations.

Prinz doesn't quite want to identify such networks with concepts for the reason alluded to above: concepts are involved in occurrent mental states that are located in working memory. When one employs a concept, an element of a relevant network is activated. That is to say, an element is tokened in working memory. When this happens, an element of the network goes proxy for the category in working memory. For example, whenever you employ the concept DOG in thought, an element of a complex network stored in your long-term memory will be tokened in your short-term memory. On different occasions and in different contexts you might token different elements of the complex. On all such occasions you are thinking a thought involving the concept DOG because the representation you token is drawn from one and the same complex, a complex that was constructed on the basis of interactions with dogs.

I began by stating that Prinz identifies concepts with proxytypes. We are now in a position to understand what this claim comes to. A proxytype is any element of a complex representational network stored in long-term memory corresponding to a particular category that could be tokened in working memory to go proxy for that category. As Prinz puts it: 'concepts are mental representations of categories that are or *can be* activated in working memory. I call these representations "proxytypes," because they stand in as proxies for the categories they represent' (2002: 149). I also began by stating that Prinz's theory is an empiricist theory, and we are now in a position to see what that claim comes to. In connection with concepts, empiricism is often characterized as the view that all our concepts are learned as opposed to being innate. Now Prinz does think that the networks to which proxytypes belong are constructed on the basis of experience and so are not part of our innate endowment. However, the representational primitives out of which they are constructed are innate. What makes Prinz's theory empiricist is that these primitives are perceptual representations, so that concepts are constructed out of perceptual resources. In other words, Prinz is endorsing Locke's ([1689] 1975) slogan that nothing is in the mind unless it was first in the senses.

There are several further features of Prinz's account that are worth bringing out. First, in virtue of the fact that different proxytypes are utilized on different occasions when thinking thoughts involving the concept DOG, we don't have a single concept DOG; rather we have many DOG concepts. However, Prinz points out, there is likely to be a default proxytype which is 'the representation that one would token if one were asked to consider a category without being given a context' (2002: 154). Second, Prinz is committed to an atomistic view of content. What gives a given proxytype its content is a matter of the content of the complex network from which it is drawn, and the content of that network is a matter of the identity of the things from which it was constructed on the basis of perceiving. For example, a DOG proxytype is an element of a network that was constructed on the basis of perceptual interactions with dogs.

A third additional feature of the account relates to Prinz's emphasis on the importance of concepts for categorization and inference. When one categorizes something as a dog, what happens is that a match is found between a current perceptual state and one of one's DOG proxytypes. And when one infers from this that the animal so categorized barks, the proxytype tokened in categorization causes the tokening of another proxytype belonging to the network that represents the barking aspect of dog behaviour. This second proxytype will have been added to the network as a result of hearing dogs bark.

At this point it should be clear that there are considerable differences between Prinz's proxytype theory and Fodor's theory, notwithstanding the fact that both are committed to an atomistic view of the content of concepts. Examining these differences can form the basis of a critical evaluation of both theories. First, for Fodor, concepts are amodal representations. That is to say, they are arbitrary symbols that do not take the form of any representations involved in perception. Prinz, on the other hand, views concepts as being built from perceptual representations that are associated with a range of modalities, so that concepts are multi-modal representations. Second, Fodor regards most lexical concepts[16] as being simple representations, whereas for Prinz such concepts are complex representations. Fodor doesn't deny that there are complex representational structures associated with concepts expressed by means of simple symbols of LOT. Consider DOG, for example. For Fodor, the fact that dogs reliably cause the tokening of this LOT symbol – thereby playing a role in fixing its content – could depend upon complex structures that represent various properties of dogs, including those that are readily perceivable. Such structures would serve as mechanisms that mediate the causal connection between dogs and DOG, but they are not to be identified with the concept DOG.

Prinz (2005) argues that identifying concepts with amodal symbols fails to explain how we categorize the things with which we interact and that this is a major failing given that categorization is one of the primary functions of concepts. Consequently, in order to make sense of categorization, Fodor also needs to postulate complex representational structures that mediate the causal connection between concepts and the items that fall under them. In the case of DOG, this complex structure will represent the perceivable properties that dogs typically have. But, Prinz continues, the upshot of this is that his account should be preferred to Fodor's on grounds of simplicity. For, by identifying concepts with the kinds of structures that Fodor regards as mediating mechanisms, he abandons any need to postulate additional amodal symbols.

A problem with this objection is that it overlooks the chief motivation for postulating the existence of a language of thought made up of amodal symbols. For Prinz, categorization involves the activation of a component of a complex network stored in long-term memory. For example, suppose I am confronted by a dog. A match is found between the perceptual state that the dog causes and a component of the network built on the basis of perceptual interactions with dogs. Thus, that proxytype is activated, an event that constitutes my categorizing the animal before me as a dog. Suppose that the dog is silent when I perceive it but that I go on to infer that it barks. This will involve the proxytype I token causing the activation of another element of the network. This element will be a proxytype that was added to the network on the basis of experiences of dogs barking. This picture of categorization and reasoning compares quite closely to that advanced by advocates of the prototype theory.

The kind of reasoning portrayed here is based upon associative learning and involves the tokening of quite simple thoughts. Thus, on seeing a dog I think DOG (or IT'S A DOG) and go on to conclude BARKS (or IT BARKS). Now perhaps the proxytype theory can handle this kind of reasoning. But much of our reasoning is far more complex than this in the respect that it involves many steps, drawing upon information from a range of very different domains, making connections which outstrip one's experience and tokening thoughts containing many concepts. Consider an example. Suppose that I have to collect my children from school by 6.00 p.m. I'm running late, as it is 5.00 p.m. and I've just come out of a meeting on a campus 30 miles away. Following my normal route home takes me fifty minutes, but I don't automatically select this route as I reason that, given the current time, that route may well be subject to traffic congestion that would slow me down considerably. So I begin reflecting in order to work out if there are any alternative routes that will get me to my destination on time. In doing this I take into account a range of factors, such as route lengths, speed limits, the number of roundabouts and junctions, the proximity of the routes to large residential areas, the amount of fuel I have in my petrol tank, and so on. I eventually settle on a route different to my normal one and arrive with five minutes to spare. This is an example of everyday reasoning, but it does seem quite distant from the kind that the proxytype theory seems well suited to handle. The relevant point in this context is that it is the kind of reasoning that has a logical character and so is readily explained in terms of the employment of logical rules or principles. But employing such rules involves applying them to representations that have an appropriate logical form. Now the simple symbols of LOT postulated by Fodor belong to a language that has syntactic rules for combining those symbols to create more complex structures. These complex structures do not merely include complex concepts such as BROWN DOG but thoughts such as THE BROWN DOG THAT LIVES NEXT DOOR INVARIABLY BARKS WHEN THE POSTMAN DELIVERS A LETTER. In other words, they include thoughts that have precisely the kind of logical forms that enable them to figure in processes of logical inference – processes that involve the application of logical rules and principles. In short, then, an important motivation for postulating amodal symbols and identifying them with concepts is to make sense of our complex reasoning capacities.

Prinz does think that proxytypes can be combined, but the kinds of examples on which he focuses involve the combination of two concepts, such as BROWN and DOG to form the complex BROWN DOG. But what he needs to show is that the proxytype theory can make sense of how we combine our concepts to create the kind of thoughts we routinely have and that the resultant structures have a form that enables them to figure in processes of logical reasoning.

In a nutshell, I have objected that Prinz focuses on simple inferences that, perhaps, can be handled by the proxytype theory but overlooks the more complex thought processes that Fodor's approach is designed to handle. For what a theory of concepts needs to do is to explain how our concepts can be combined to form the complex thoughts we are capable of having and to do so in such a way that makes clear how such thoughts could figure in the reasoning processes we routinely execute.

A second objection to the proxytype theory relates to abstract concepts. Many of the concepts we possess are abstract, in the sense that they either refer to properties that cannot be directly perceived or group together phenomena that diverge widely in their perceivable properties. We saw that this posed a problem for the imagistic theory, as many concepts do not seem to be such that their contents are readily expressed by means of an image. Concepts such as those of FAIRNESS, TRUTH and CAUSATION are prominent features of our mental lives, but what exactly does fairness, truth or causation look like? The potential problem for Prinz is that he thinks that the representational primitives out of which proxytypes are built are perceptual – that is, they are the representations employed by our perceptual systems. But, one might reasonably ask, how can abstract concepts be built out of such non-abstract resources?

Prinz does not shy away from this problem. He begins by pointing out that it is far from clear that anti-empiricists who regard concepts as being amodal symbols have any advantage with respect to abstract concepts. Merely postulating an amodal representation and assigning to it an abstract content does not thereby explain how that representation could come to have that content. Prinz points out that the standard form of explanation of the content of abstract concepts developed by friends of amodal representations appeals ultimately to causal factors. For example, our concept of FAIRNESS represents fairness because its tokenings are reliably caused by instances of fairness – that is, people, events and situations that are fair. But, argues Prinz, such a reliable causal relation involves perception and perceptual representation as we perceive individuals, events and situations that are fair. Moreover, it requires a correlation between fairness and readily perceivable features of those people, events and situations that are fair. In other words, the friend of amodal concepts is committed to the possibility of tracking or detecting abstract properties via concrete properties that are perceivable and so represented by means of perceptual representations. Prinz attempts to take advantage of this aspect of the causal approach to content. He argues that our perceptual representations are capable of expressing abstract properties because they track or detect their presence via their correlation with concrete properties.

How successful is this attempt to deal with abstract concepts? I think Prinz is right to suggest that it is not just friends of perceptual representations who have a problem accounting for abstract concepts. Nevertheless, I have doubts about the viability of appealing to tracking and detecting by causal means. Suppose that a particular abstract property is correlated with concrete ones so that someone in the know could perceptually detect the presence of that abstract property. But that is going to work only if they have a representation to express the abstract property that is distinct from those that express the correlated concrete properties. If there is no such distinction, then the representations reliably caused by instances of the abstract property will serve to express the concrete properties of the things, events and situations perceived. In other words, no abstract concepts will be involved. But if Prinz does postulate representations independent of the perceptual ones that are involved when we detect the presence of abstract properties, then he is in danger of abandoning his view that all concepts are – or are constructed out of – perceptual representations. In other words, Prinz's problem is that of explaining how any of our concepts can represent abstract properties *as such* rather than the concrete properties that are correlated with those abstract properties.

A third objection to the proxytype theory relates to Prinz's account of how proxytypes get their content. Prinz argues that the DOG proxytypes have the content they have because they are drawn from a complex network that was built on the basis of interactions with dogs. This readily accounts for misrepresentation, for if, say, a fox causes the tokening of a proxytype from this network, the fox will have been misrepresented as a dog in virtue of the historical origins of the proxytype. However, Prinz also argues that the networks are constructed over time and that at any point in their history new elements can be added to them. For example, if I encounter a Pomeranian for the first time, I may well add more to the DOG network in order to reflect what is distinctive about Pomeranians. But this generates a problem, for it is highly likely that at some point interactions with non-dogs has led to additions to the putative DOG network, implying that that network was constructed on the basis of interactions with a category of creatures broader than that of dogs, with the implication that proxytypes drawn from that network have a content broader than *dog*.

A final problem for the proxytype theory relates to a widely accepted idea motivated by Hilary Putnam's (1975) Twin Earth thought experiment. Putnam argued that the meaning of words such as 'water' on our lips is not exhaustively determined by our intrinsic physical states; in addition, the nature of the external environment plays a role in fixing meaning. In order to support this claim, Putnam constructed the following thought experiment. Oscar is a normal resident of Earth who routinely applies the word 'water' to the odourless, colourless stuff that falls as rain, fills rivers and lakes, comes out of taps, and so on. As that stuff is H_2O, Oscar's word 'water' correctly applies to, and only to, H_2O. Twin Oscar is a molecule-for-molecule duplicate of Oscar who lives on Twin Earth, a planet just like Earth apart from the fact that the odourless, colourless stuff that falls as rain, fills rivers and lakes, comes out of taps, and is routinely called 'water' is XYZ rather than H_2O. Consequently, Twin Oscar's word 'water' correctly applies to, and only to, XYZ. As what a word correctly applies to – its reference or extension – is part of its meaning, the word 'water' on Oscar and

Twin Oscar's respective lips diverges in meaning, despite the fact that the twins are physically indistinguishable.

Putnam's argument about linguistic meaning has been extended by many philosophers to make a parallel point about concepts and the thoughts in which they figure.[17] Thus, so the argument might run, the concept that Oscar expresses by means of the word 'water' and the thoughts that he expresses by means of sentences such as 'water is wet' have a different content than the corresponding concept and thoughts of Twin Oscar. Therefore, the content of our concepts and thoughts is at least partly determined by the nature of the environment we inhabit. This view is known as externalism and has been widely endorsed in the philosophical community.[18]

Prinz himself endorses externalism, and this presents him with a problem, namely that of explaining how Oscar on Earth and Twin Oscar on Twin Earth could have learned divergent concepts. More specifically, how could the concept that Oscar comes to express by means of 'water' diverge in content from that which Twin Oscar expresses by the same word? The problem for Prinz is that, given that the representational structures stored in long-term memory from which proxytypes are drawn are ultimately constructed out of perceptual representations, it would appear that the twins have exactly the same proxytypes and, therefore, exactly the same concepts.

Prinz is alive to this problem, and in addressing it he employs Locke's distinction between real and nominal essences. The real essence of water (that is, the colourless, odourless liquid found here on Earth) is a matter of its microphysical constitution. The nominal essence of water is a matter of the perceivable properties characteristic of water on the basis of which we typically identify a sample of water as such. Corresponding to this distinction is that between real and nominal content. The real content of the respective concepts expressed by means of 'water' by Oscar and Twin Oscar differ. This is because the stuff falling under Oscar's concept has the real essence of being H_2O while the stuff falling under Twin Oscar's concept has the real essence of being XYZ. On the other hand, their concepts have the same nominal content as the perceptual representations that figure in the proxytypes that constitute their respective concepts. In effect, what Prinz is saying is that the real content of a particular concept possessed by an individual is a matter of the essence of the items with which the individual causally interacted in constructing that concept. As Oscar interacted with H_2O in constructing his concept, that concept has the real content *water*, whereas Twin Oscar's corresponding concept has the real content *twin water*, as it was constructed on the basis of causal interactions with Twin Water.

However, what I have said so far leaves out a crucial aspect of Prinz's line of thought, and this is his endorsement of psychological essentialism, the doctrine mentioned above in the context of the theory theory of concepts. Thus, with respect to Oscar, Prinz would say that he thinks of the stuff falling under his concept WATER as having a particular essence (the nature of which he may well think himself ignorant) that is the causal basis of the perceivable properties in virtue of which he typically identifies a sample of water as such (that is, the properties that are represented by the relevant proxytype). Thus, Prinz accounts for the real content of Oscar's (and our) concept WATER on the basis of Oscar's (and our) essentialist commitments, along with the fact that that concept was constructed on the basis of causal interactions with H_2O. Without such an essentialist commitment, the concept expressed by Oscar (and us) by means of 'water' would have a content such as to apply to anything with an appearance like that of water. Thus, it would apply to XYZ as much as to H_2O.

I'm not convinced by this line of thought for the following reasons. Not all types of stuff for which we have concepts have the same type of essence. Water, for example, does have a microphysical essence; a sample of stuff is a sample of water if and only if it is composed of H_2O molecules. Milk, on the other hand, does not have a microphysical essence. Suppose that a super-intelligent robot on a distant planet synthesized two samples of liquid, one physically just like the milk in my fridge and the other physically just like the water in the bottle from which I have recently taken a sip. In neither case were the liquids drunk by any animal and in neither case were they synthesized in order to be drunk or so benefit health. My contention is that the water-like substance synthesized by the robot is indeed water, whereas the milk-like substance is not milk because it was neither produced within the body of an animal nor used by an animal to support its growth and development. Moreover, the stuff that they call 'milk' on Twin Earth is indeed milk, even though it is made up largely of XYZ, as that stuff has the relevant origins and function. In sum, milk has a bio-functional essence, an essence relating to its biological origins and use in supporting health and growth, rather than a microstructural essence.

An upshot of this is that, if a commitment to essentialism is going to help us to learn the concepts we acquire on the basis of our experiences, it had better not take the form of an unarticulated, generic notion of essence. For such a generic notion of essence won't serve to pin down the relevant type of essence. For example, if a child interacting with milk doesn't think of the white stuff she sees as having a bio-functional essence, then she won't acquire the concept MILK. Rather, she will acquire a concept with an indeterminate content – that is, a content indeterminate between that of the concept MILK and that of a concept that applies to, and only to, stuff with the same microphysical structure as the milk she has experienced. With respect to water, what she will need to do is think of the stuff with which she is interacting as having a microphysical essence rather than one relating to its origins and role in sustaining the health of living things. In short, if a commitment to essentialism is going to help children learn concepts in the manner that Prinz suggests, then they will need to have a range of different understandings of what an essence can amount to and employ the relevant understanding in each specific case.

Now it might be argued that children do indeed have a range of different notions of essence that they can employ in learning concepts. However, that raises the difficult question of how they learned such a range of notions of essence and is in tension with the empirical evidence suggesting that a typical child has a 'placeholder' notion of essence, as noted above. Moreover, as Bloom (2004) points out, not even highly educated parents talk to their children about essences, so children are not typically in receipt of explicit guidance on these matters.

It follows from these reflections that Prinz cannot explain how Oscar and Twin Oscar can diverge in their concepts. Given that the twins also acquire concepts that apply to types of stuff that do not have microphysical essences (for example, MILK), they cannot acquire concepts such as WATER or TWIN WATER by using a generic, unarticulated notion of essence. Rather, they must each have a range of different notions of essence and apply the appropriate one in each particular case. For example, Oscar must apply a notion of microphysical essence when interacting with water and a notion of bio-functional essence when interacting with milk, otherwise he will not acquire the concepts WATER and MILK on the basis of those respective interactions. But this raises the question of how Oscar could have acquired such notions of essence and know which to apply in each particular case. Prinz, I contend, has no answer to these questions.[19]

7 Conclusion

In this chapter I have examined the most prominent contemporary theories of concepts. These are the prototype theory, the exemplar theory, the theory theory, informational atomism and the proxytype theory. I have argued that, despite their respective strengths and the ingenuity that their advocates have exercised in defending them, none is without its problems. Consequently, there is still much work to be done in constructing a viable theory of the nature of human concepts.

5 Language

1 Introduction

Language plays an important role in our cognitive lives. It is the primary means by which we express our thoughts to our fellows and enables us to communicate complex messages efficiently and effectively. Some philosophers[1] have argued that thought is impossible without language. But even if we reject this claim it is difficult to deny that our thinking would be impoverished without language for at least two reasons. First, shared mastery of language enables knowledge to be passed from one individual to another, so that each individual can benefit from the cognitive endeavours and successes of their fellows. Second, mastery of language enables us to frame our thoughts to ourselves in a way that makes them accessible to consciousness and easier to store and retrieve from memory (Jackendoff, 2002, 2012). In virtue of this, language makes us better thinkers.

Given this, it should come as no surprise that language has been a key concern within cognitive science. With respect to language, two questions stand out, and in this chapter I will examine some of the key debates surrounding those questions. The first question concerns the cognitive basis of language: when an individual has mastery of a language, in virtue of what cognitive structures do they have that mastery? The second question is this: how do we acquire the cognitive structures that ground our linguistic capacities? For example, are they learned, innate or a mixture of both?

2 Knowledge of language

A natural way to think of the cognitive basis of an individual's mastery of language is in terms of knowledge. For example, if an individual has a mastery of English, then this is based upon a rich body of knowledge about the English language on which they draw in speaking and understanding English. Such knowledge will include knowledge of what words belong to English, their meaning and their sound (how they are pronounced), and of how words can be combined to create larger linguistic units such as phrases and sentences. Invoking knowledge of language in this way enables us to follow Noam Chomsky (1986) in expressing the above two questions in the following manner:

- What does an individual know when they have mastery of a language?
- How do they acquire that knowledge?

Chomsky's seminal work in the tradition of generative transformational grammar has provided the most prominent answer to these questions, and in this chapter I will focus on that work and the heated debate that has surrounded it, thereby introducing some alternative perspectives.

Beginning in the late 1950s with his short book *Syntactic Structures* (1957) and his blistering review of Skinner's *Verbal Behavior* (1959), Noam Chomsky revolutionized linguistics, the scientific study of language. He overthrew the then dominant structuralist tradition, exemplified by the work of Leonard Bloomfield (1933), with its behaviourist leanings and its conception of language as a collection of sentences spoken in a speech community. This he replaced with an approach known as generative transformational grammar, an approach that has dominated linguistics ever since. The views of Chomsky and his followers have changed over the years, but his overall perspective has remained remarkably consistent. In its present incarnation Chomsky's perspective is known as minimalism (Chomsky, 1995; Boeckx, 2006), which is itself a development of the principles and parameters approach (Chomsky, 1981).[2] Chomsky's work can be very technical and forbidding, but a relatively easy way into his outlook can be achieved by contrasting it with a very commonsensical view of language that can be described in the following terms.

Languages such as English are social entities. They were created collectively by groups of individuals to facilitate communication between themselves and their fellows. Thus, what words belong to the language (what those words mean and how they sound) and what rules govern how those words can be combined (the grammatical or syntactic rules of the language) are matters of social convention. Thus, languages are akin to other rule-governed social structures such as games like chess, legal systems and codes of etiquette. Accordingly, they are not set in aspic but can change over time as they are passed on from one generation to the next. Moreover, just as different social groups can play very different games, establish divergent legal systems and follow conflicting rules of etiquette, they can speak radically different languages.

In order to speak a given language an individual must be in an appropriate psychological state; she must have a rich body of knowledge about that language. That is, she must have knowledge of some of the words that belong to the language and how those words can legitimately be combined, and such like. Such knowledge is about a particular complex social entity and is acquired by means of learning based upon the employment of general learning mechanisms. An individual's knowledge of language need not be complete and can come in degrees. For example, a five-year-old child or an adult learning English as a second language will have a body of knowledge about English that is less extensive than that of a typical reader of this book. But even such an individual wouldn't have a complete knowledge of English. Indeed, an individual will often be mistaken about some aspect of the language she speaks, as when, for example, she believes that a particular word means something that it doesn't or that a particular structure is grammatical when it is not. This is so in virtue of the fact that knowledge claims about a particular language are answerable to social facts that lie outside of the mind of any speaker of that language. Thus, languages have a substantial normative dimension in that the distinctions between right and wrong, correct and incorrect, use and misuse very much apply.

An upshot of all this is that the study of language is a branch of the social sciences concerned with the social structures with which individual speakers of the target language are involved and to which they are answerable. Although the social structures that constitute English, for example, would not exist without individual speakers of English, they are nevertheless external to any such individual speaker.

The conception of language I have been describing over the previous three paragraphs is not merely the commonsense view but is widely held in philosophy, linguistics and cognitive science more broadly.[3] It is the view that Chomsky is vociferously opposed to. For him language is a psychological entity – that is, it exists in the mind of the speaker and nowhere else. In *Aspects of the Theory of Syntax* (1965) Chomsky argued that, in virtue of this, linguistics is a branch of cognitive psychology. In *Knowledge of Language* (Chomsky, 1986) he introduced the distinction between I-Language and E-Language or between I-Language conceptions of language and E-Language conceptions of language. The 'I' in 'I-Language' stands for 'internal', 'intrinsic' and 'intensional' and the 'E' in 'E-Language' stands for 'external', 'extrinsic' and 'extensional'. Thus, according to advocates of the E-Language conception, languages are external and extrinsic to any individual speaker. And, according to the I-Language conception, languages are internal and intrinsic states of the minds of speakers. Chomsky endorses the I-Language conception and rejects the E-Language one, at least for the purposes of the scientific study of language. Thus, he describes his work as involving a 'shift in focus . . . from the study of language as an externalized object to the study of the system of knowledge of language attained and internally represented in the mind/brain' (1986: 24).

More recently the term 'biolinguistics' has been used to characterize Chomsky's approach (Jenkins, 2000). One reason for this is that he thinks that the mind is ultimately the brain, so that to describe an individual's language is to describe a state of their brain, a biological system, at a relatively high level of abstraction. Thus, Chomsky writes:

> for H to know the language L is for H's mind/brain to be in a certain state; more narrowly, for the language faculty, one module of this system, to be in a certain state S_L. One task of the brain sciences, then, is to discover the mechanisms that are the physical realization of the state S_L. (1986: 22)

A second reason relates to his view of language acquisition, something that I will come to shortly, but first I will explore a consequence of his I-Language perspective.

Advocates of the commonsense view described above typically think that, in order for an individual to speak a particular language, they must be in an appropriate psychological state – namely, one that serves to connect them to the language in question. Such a state could represent the language in question or constitute a body of knowledge about it.[4] Now Chomsky has described speakers of a language as knowing that language (Chomsky, 1986) and of having a representation of it encoded in their mind (Chomsky, 1980). However, one has to be careful how one understands these terms as, for him, there is nothing external to a speaker's body of knowledge of their language that constitutes their language and that their knowledge is about. Similarly, the representations that encode an individual's language are not representations of anything external to them. In other words, the knowledge and representations *are* the individual's language rather than something that is directed at, and serves to connect them to, some external social entity that constitutes their language.

Now suppose we ask how many languages there are. In one respect Chomsky would be happy with the answer that there are as many languages as there are individual speakers (or indeed, time slices of individual speakers). This is because there is always likely to be some difference between the psychological states relating to language between any two individuals, even if they would ordinarily be characterized as speaking the same language. Hence, if their language is a matter of their psychological state then they speak two different languages, though in many cases those languages will be sufficiently similar to facilitate reliable communication. On the other hand, Chomsky would also be happy with the answer that there is only one human language, on the grounds that the differences between the language constituting psychological states of any two individuals are very limited and would be regarded as irrelevant by a visiting super-intelligent Martian. This point brings us to Chomsky's views of language acquisition.

It is uncontentious that there are distinct internal organs of the human body – the heart, lungs, liver, and so on – each of which has its own distinctive function but is able to serve that function only in virtue of its interactions with other organs. Such organs are part of the human biological endowment. They are subject to growth and development from their appearance *in utero* to their mature state in the adult individual. Such a process relies upon appropriate environmental conditions. For example, the development of healthy adult lungs requires that the individual has an adequate diet and avoids the regular inhalation of noxious fumes. Nevertheless, there is a natural developmental trajectory followed by internal organs that is constrained and directed by our common, shared human genetic makeup.

For Chomsky, the mind, or, as he calls it, the mind-brain, consists of functionally discrete yet interacting components that are akin to the internal organs of the body. In other words, he is an advocate of a modularity thesis.[5] One such component relates to language, and he calls it variously the language organ or language faculty. This language faculty is part of our human biological endowment and undergoes a process of development that, though requiring appropriate environmental conditions and stimulation, is constrained and directed by our common, shared human genetic makeup. Such a process of development involves moving from an initial state at birth to a stable, mature state. Such stable, mature states *are* languages in the respect that an individual's language just is the stable, mature state of her language faculty. Such states enable individuals to produce and understand sentences of their language. The process of moving from the initial state to the mature state is not one that is mediated by general learning mechanisms. As Chomsky puts it:

> knowledge of a particular language grows and matures along a course that is in part intrinsically determined, with modifications reflecting observed usage, rather in the manner of the visual system or other bodily 'organs' that develop along a course determined by genetic instructions under the triggering and shaping effects of environmental factors. (1986: 2)

And again:

> Language acquisition seems much like the growth of organs generally; it is something that happens to a child, not that the child does. And while the environment plainly matters, the general course of development and the basic features of what emerges are pre-determined by the initial state. But the initial state is a common human possession. It must be, then, that in their essential properties and even down to fine detail, languages are cast in the same mold. The Martian scientist might reasonably conclude that there is a single human language, with differences only at the margins. (2000: 7)

One key difference between the language faculty and internal bodily organs that has already been alluded to is that they are, as Chomsky (2000) puts it, 'computational-representational' systems. That is to say, the state of an individual's language faculty represents or encodes their language and that state underlies their acts of producing and understanding sentences. Thus, for example, if my language faculty were in a very different state from its current state (say, by being in one like that of Pierre, a monolingual French speaker), then that would be reflected in what sentences I produced in expressing my thoughts and in how I understood sentences that I heard other people speak.

The state of an individual's language faculty (that is, their language or, better, their I-Language) is often referred to as a 'grammar'. Grammars themselves have different elements and so are modular in structure. Two of the most important elements of a grammar are a lexicon and a collection of syntactic rules for combining items belonging to the lexicon into larger linguistic structures such as phrases and sentences. A lexicon is, in effect, a vocabulary. It contains all the 'words' that an individual knows and represents information that is specific to them and distinguishes them from other words. This information includes their sound (that is, how they are pronounced), their meaning and which syntactic category they belong to.

When we speak we rarely produce single words; typically, we produce a sentence. Thus, in a language, words can be combined to produce larger linguistic structures. But not any combination of words is a legitimate structure in a language. For example, 'the dog will chase the postman' is a legitimate sentence of English but 'chase postman dog the the' is not, even though it contains just the same words. What determines how words can be combined in a language, and thus which structures are legitimate and which not, are the syntactic rules of the language. Hence, an important part of an individual's grammar is a collection of syntactic rules. These can be thought of as rules for building complex linguistic structures out of words.

An upshot of all this is that the central task of the linguist concerned with, for example, English would be to describe the grammar encoded in the language faculty of a typical or idealized speaker of English. But what, one might ask, do the crucial syntactic rules look like? An important point to note is that they are unconscious and so not revealed by introspection. A second point to note is that, though they are finite in number, these rules generate infinitely many distinct sentences. The term 'generate' is to be understood in the logico-mathematical sense of implying or entailing rather than in the mechanical sense of concretely producing. These two notions come apart, as one could have a structure that is a legitimate sentence of a language even though no speaker ever has produced or ever could, in practice, produce it, because, for example, it was too long. This naturally leads to the famous competence/performance distinction (Chomsky, 1965). An individual's linguistic competence is her knowledge of her language, which, as we have seen, is a state of her language faculty. Her performance has to do with her actual and potential linguistic behaviour. Although competence underlies performance in the respect that an individual draws on her knowledge of language when engaging in linguistic behaviour and wouldn't be able to so behave without such knowledge, the two are distinct. An individual could have knowledge of a language even though she couldn't speak that language or manifest that knowledge due to, say, brain damage. And, even in normal cases, our linguistic behaviour does not perfectly reflect our competence, as it is influenced by factors external to the language faculty such as memory and attention limitations (Chomsky, 1980).

A third point relates to what Chomsky calls linguistic creativity. Individuals routinely produce – with full understanding – and routinely hear – and fully understand – sentences they have never encountered before. This is something which requires explanation. According to Chomsky, what explains this creativity is that syntactic rules are recursive in nature. What this means is that rules can be applied to their own output. That is, when certain rules are applied to produce a particular structure, those rules can be applied once more to that structure to create a more complex structure, and so on. For example, the rule that was used to construct the noun phrase 'black dog' in the sentence 'the black dog chased the postman' can be applied to that phrase to construct the noun phrase 'hairy black dog'. And it can be applied to that noun phrase to construct the noun phrase 'ferocious hairy black dog', and then 'smelly ferocious hairy black dog', and so on. A second example relates to sentences containing psychological verbs such as 'think'. The sentence 'Mark thinks that Paris is beautiful in the spring' can be embedded in the more complex sentence 'Steve thinks that Mark thinks that Paris is beautiful in the spring'. This can be embedded in the yet more complex 'Dan thinks that Steve thinks that Mark thinks that Paris is beautiful in the spring', which, in turn, can be embedded in the even more complex 'Constantine thinks that Dan thinks that Steve thinks that Mark thinks that Paris is beautiful in the spring', and so on. In other words, as a result of the recursive nature of the rules, there is no limit to the number of distinct sentences in a language such as English, even though it has only finitely many words and finitely many rules for combining those words. Therefore, an individual who has knowledge of such rules and a collection of words has at her disposal the means to construct and understand countless distinct sentences.

A fourth point is that syntactic rules are such as to generate linguistic structures that are hierarchical in nature rather than linear collections of words. The hierarchical nature of sentences is captured in the tree diagrams used by linguists to lay bare the internal structure of sentences. In constructing a sentence, words are combined to create a phrase. That phrase is in turn combined with another word or phrase to create a higher-level phrase, which is in turn combined with a word or phrase to create a yet higher-level phrase, and so on. This combining operation is known as merger, and it is a binary operation that always merges just two elements.

A phrase always has a head. This is, so to speak, the most significant constituent word of the phrase, and the syntactic category to which it belongs determines what type of phrase the phrase is. For example, if the head of a phrase is a noun, then the phrase is a noun phrase; if the head of a phrase is a verb, then the phrase is a verb phrase; and so on. Now consider the following sentence:

The postman will chase the dog.

This sentence can be derived in the following manner. The determiner 'the' is merged with the noun 'postman' to form the determiner phrase 'the postman'. That noun phrase is merged with the verb 'chase' to form the verb phrase 'chase the dog', which is then merged with the auxiliary verb 'will' to create the T' (pronounced T-bar) 'will chase the dog'. 'Will' is a tense constituent, as it indicates that the phrase it heads describes a future event. The T' is a level of structure lying between a basic constituent (in effect, a word) and a full phrase.[6] This T' is then merged with the determiner phrase 'the postman' (itself formed from the merger of 'the' and 'postman') to make the tense phrase 'the postman will chase the dog'. Finally, this tense phrase is merged with a complementizer to make a complementizer phrase.[7] A complementizer is a constituent of a structure that indicates its force – that is, whether it is, for example, an assertion, a question or a conditional. Familiar examples include 'if' in 'if it is sunny tomorrow I will go for a long run' and 'for' in 'for me to take a day off work it will have to be sunny'. In the present case, the complementizer is the null assertoric complementizer. That is, it indicates that the structure serves to make an assertion but it is not vocalized. As we shall see, Chomsky thinks that linguistic structures often contain null (that is, non-vocalized) constituents.

The hierarchical structure of 'the postman will chase the dog' can be represented by a tree diagram of the kind much loved by linguists (see figure 5.1).

In addition to rules for merging constituents and phrases, there are rules for moving elements in structures to create a new structure. One prominent example of such a type of rule is that involved in the formation of passive from active sentences (for example, 'the dog was chased by the postman' from 'the postman chased the dog'). Another is that involved in 'wh-movement', when a 'wh-word' (such as 'who', 'what', 'when', and so on) is moved to form a question (for example, 'what did the postman chase?' from 'the postman chased what?').

Producing a sentence thus involves taking items from the lexicon and merging them and moving elements around in accord with rules encoded in the language faculty so as to create a hierarchically organized structure.

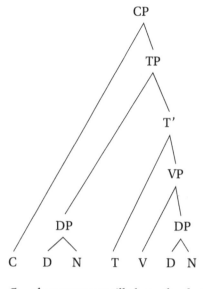

Ø   the postman will chase the dog

Figure 5.1

That structure is represented in the language faculty and is delivered to adjacent performance systems concerned with articulation. Similarly, understanding a sentence produced by someone else involves working out the identity of its basic constituents and how they have been put together by application of the merge and movement rules. Once again, this involves the construction of a representation akin to a tree diagram, which, in this case, is delivered to adjacent performance systems involved in thinking. The upshot of this is that the internal structure of the sentences we vocalize is determined by the underlying representations in the language faculty.

One further advantage of postulating such internal representations is that it enables us to explain ambiguity in those cases where a sentence has two possible meanings due to its having two possible syntactic structures. Consider the sentence 'the man can see the boy with the telescope'. This can mean either that the man has the telescope and is using it to see the boy or that the boy has the telescope. These two possible structures can be represented as in figures 5.2 and 5.3, and what a person means in producing the sentence (or understands in hearing the sentence) depends on which structure is represented in their language faculty. The structure shown in figure 5.2 means that the man has the telescope, in contrast with that in figure 5.3, which means that the boy has the telescope.

One very important syntactic relationship emphasized by Chomsky's work within its minimalist phase is that of constituent-command (c-command for short). Suppose that two constituents (be they words or

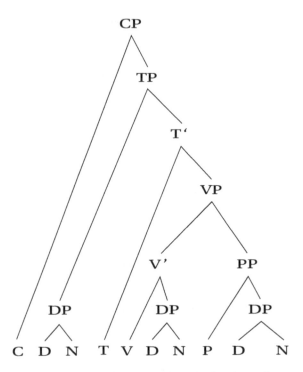

Figure 5.2

phrases) are merged. Then they are sisters of each other. They are also both contained within the resulting structure. Thus, in the DP 'the dog', the determiner 'the' and the noun 'dog' are sisters of each other and are both contained in the DP. As Radford (2009: 59) defines it: 'A constituent X c-commands its sister constituent Y and any constituent Z which is contained within Y.' So, for example, in the sentence 'the dog will chase the postman', the verb 'chase' c-commands 'the postman' and its components 'the' and 'postman'. 'The', on the other hand, c-commands only 'postman' and 'postman' c-commands only 'the'.

C-command is invoked to explain a range of syntactic phenomena. One example relates to anaphors or reflexive pronouns such as 'himself', 'herself', 'themselves', and so on. Such words don't refer directly to entities outside of the sentence but, rather, take their reference from an appropriate antecedent referring expression in the phrase or sentence. The rule relating to anaphors is that they must be c-commanded by an appropriate antecedent. This condition is satisfied in the case of the sentences 'Bill shot himself in the foot' and 'she is ashamed of herself'. In the first of these sentences, the noun 'Bill' c-commands 'himself' and in the second the pronoun 'she' c-commands

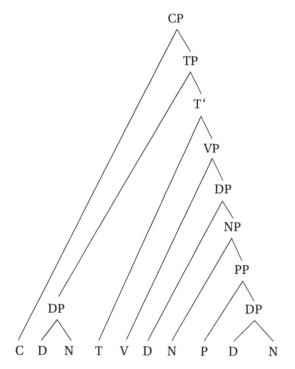

Ø the man can see the boy with the telescope

Figure 5.3

'herself'. However, it is not satisfied in the case of 'himself saw Bill' and 'she is fond of himself'. In the former sentence, 'himself' is not c-commanded by any antecedent referring expression. In the latter, although 'she' c-commands 'himself', these two pronouns do not agree in gender.

I claimed that, for Chomsky, sentences often contain null (unvocalized) constituents. Another example is PRO, sometimes known as 'big pro', which is the null subject of some infinitive complement clauses. Consider the following sentence:

> I want [to go home].

The element in brackets is itself a clause that is embedded in a larger clause.[8] This embedded clause is infinitive, as its main verb is in the non-inflected infinitive form. This clause doesn't have an overt subject, in contrast to the complement clause in:

> I want [him to go home].

Yet it is clearly about me; the sentence as a whole is saying that I want me to go home. Hence, its subject is an unvocalized pronoun-like element that takes its reference from the subject of the main clause. This is known as PRO (big pro), so that a better characterization of the sentence as it is represented in the language faculty when produced or understood is:

I want [PRO to go home].

From the perspective of contemporary generative grammar, the analysis of such sentences as 'the dog will chase the postman' provided in this section have been heavily simplified. For example, I have made no mention of the movement of 'the dog' from inside the verb phrase 'the dog chased the postman' to the specifier position in TP, a movement which involves copy and deletion, whereby a null trace of 'the dog' is left in its original position. Nor have I made any mention of the various features that the constituents carry, such as the case features carried by the DPs or the extended projection principle feature carried by 'will'. These features need to be checked and deleted via merger with an appropriate constituent (and this often requires movement) or by being the goal of an appropriate probe at a higher level in the structure. Don't worry if this linguistic terminology is somewhat opaque. The key point is that the structure of even the simplest sentence is, from Chomsky's perspective, very complex and involves the operation of highly abstract rules implicating elements and relations that are invisible to the naked ear, so to speak.[9]

3 Chomsky's account of language acquisition

If speakers of a language have knowledge of syntactic rules that can be used to generate structures of the kind described in the previous section, then the question arises of how they acquired that knowledge. The obvious answer is that we learn such rules; after all, people with English as their first language grew up hearing people around them talking English, whereas people who don't grow up in an English-speaking community acquire knowledge not of English but of some other language. However, according to Chomsky, this 'obvious' answer is not true. This is because the linguistic experiences we have when we are acquiring our first language are not rich enough to facilitate learning. Consequently, we must have a rich body of innate knowledge about language that facilitates linguistic development.

In a nutshell, Chomsky thinks that it would be very difficult for a child to learn the rules of syntax of their first language on the basis of the kinds of experiences that children typically have. Consider the game of chess. Chess players clearly learn the rules of chess. How do they do this? Explicit instruction will often play a key role: the teacher will tell the learner what the rules are. Another way of learning would involve watching a game of chess and trying to work out the rules by constructing hypotheses and testing them in the light of experience. Learning this way would involve observing an awful lot of chess play where most of the moves made were legal; it couldn't be done on the basis of observing a mere handful of moves or a body of moves peppered with illegal moves. Children do not plausibly learn the syntactic rules of their first language by means of explicit instruction: most carers are not in a position to state the relevant rules, as these are unconscious, and even if they did the children would not understand what was said to them. Could children learn in the second way? Learning the rules this way would mean hearing lots of sentences and working out what the relevant rules were. This would involve tentatively hypothesizing particular rules, testing them against further linguistic evidence, and modifying them accordingly until the correct rules were hit upon and confirmed. Chomsky argues that children do not learn syntactic rules in this way, as their linguistic experiences are far too limited or impoverished to enable them to do so. This is the so-called poverty of the stimulus argument. Actually there is more than one poverty of the stimulus argument that has been championed by followers of Chomsky. What I will describe first is the version most clearly associated with Chomsky.[10]

Suppose a child is trying to learn the syntactic rules of English. All the evidence they have to go on consists of the sentences they hear being spoken around them. This is known as the primary linguistic data (pld for short). Chomsky argues that children can't learn from the pld: the pld is consistent with many conflicting hypotheses as to what the syntactic rules of the target language might be and so doesn't provide enough evidence as to what the actual rules are. One problem is that the pld contains lots of ungrammatical sentences: 'a good deal of normal speech consists of false starts, disconnected phrases and other deviations from idealized competence' (Chomsky, 1972: 113). Another is that the pld lacks certain sentences that are crucial to learning. This suggests that children come to the language acquisition task with innate knowledge about language (that is, knowledge that they are born with and that they do not learn).

Chomsky (1972) has made the poverty of the stimulus argument concrete by focusing upon a particular example related to the formation of yes–no questions from declarative sentences. Consider the sentence:

(1a) The man is singing.

As any competent speaker of English knows, the associated yes–no question is:

(1b) Is the man singing?

What rule do we acquire and use in forming such questions? One candidate is a simple rule (R1):

(R1) Move the first occurrence of an auxiliary to the front of the sentence.

The problem with (R1) is that it will generate an ungrammatical sentence when applied to a declarative sentence such as (2a):

(2a) The man who is happy is singing.

The yes–no question corresponding to (2a) is (2c) rather than (2b).

(2b) Is the man who happy is singing?*[11]
(2c) Is the man who is happy singing?

The rule that is needed is a rule that appeals to structural features of sentences, a rule such as (R2):

(R2) Move the auxiliary immediately following the subject determiner phrase to the front of the sentence.

Chomsky argues that children are not able to learn a rule such as (R2) on the basis of their linguistic experiences. This is because the pld contains only sentences that are equally consistent with (R1). In other words, a child acquiring language never encounters complex sentences such as (2c) – sentences that would show her that (R1) was incorrect and motivate her to hypothesize and confirm a rule such as (R2). Given this, if learning was involved in the acquisition of syntactic rules, then children would alight on (R1), as it is a simpler and more intuitively obvious rule than competitors such as (R2). However, there is no evidence that children acquire the simpler rule. Once they have settled upon a rule and begin to deal with sentences more complex than those found in the pld, they never produce sentences such as (2b). Chomsky concludes from this that some innate language-specific knowledge or constraint must explain why a rule such as (R1) rather than (R2) is acquired.

If we bring innate knowledge to the task of acquiring language, what exactly is that knowledge? According to Chomsky, it is not knowledge of English, Italian or any particular language. Rather, it is knowledge of Universal Grammar (UG for short). UG consists of a collection of rules or principles[12] that are common to all languages. UG is a template for language that constrains the form that any human language can take. All languages are thus variants on a theme laid down by UG and so can differ in only relatively minor respects.

Associated with some of the principles of UG are options that are known as parameters of variation. Languages that differ syntactically take up differing options. A classic example relates to heads. As we have seen, sentences are not linear collections of words; rather, they are entities that have a hierarchical structure. They are made up of phrases, that are made up of phrases, and so on, that are ultimately made up of words. Consider the sentence:

The leader of the module failed the student.

There are several phrases contained within this sentence, all of which, as we have seen, have a head. Some of those phrases also contain a complement. For example, the complement of the head in the noun phrase 'leader of the module' is the prepositional phrase 'of the module'. The complement of the head of that phrase is the determiner phrase 'the module', the head of which, in turn, is the determiner 'the', whose complement is the noun 'module'. As these examples suggest, in English phrases, heads always go before complements, so it is a rule of English that the head of a phrase precedes its complement. Hence, English is known as a head-first language. But not all languages are head-first languages. For example, in Japanese and Turkish, heads always come after their complements, so that those languages are head-last languages.

Chomsky argues that it is a principle of UG that heads always occupy a fixed position relative to their complements. But associated with that principle is an option (a parameter) which allows any given language to be either head-first or head-last. What UG prevents is a language that 'mixes and matches', a language that, for example, has some constituent phrases where the head precedes its complement and others where the head follows its complement.

For Chomsky, we humans have innate knowledge of UG. This is part of our biological endowment, something that is genetically determined or encoded in our genes. Thus, UG is part of human nature, and it is an aspect of human nature (rather than a social or cultural invention) that we speak languages and that human languages have a particular distinctive form. The child that grows up in an English-speaking community innately knows that the language she is trying to pick up is such that the position of heads relative to their complements is uniform. Initially she doesn't know whether this language is head-first or head-last. But she needs to hear only a small number of sentences of English to find out. When she hears such sentences, the head parameter is automatically set to the 'head-first position'. This phenomenon is often compared with the flipping of a switch set in either of two positions. A child growing up in a Japanese-speaking community will have the same parameter set in the 'head-last' position on the basis of hearing a small number of Japanese phrases. Thus, acquiring knowledge of the syntactic rules of one's first language is not a matter of learning what those rules are. Rather, it is a matter of setting parameters associated with principles of UG on the basis of a limited exposure to a particular language.

A second version of the poverty of the stimulus argument is known as the logical problem of language acquisition (Pinker, 1989). The problem relates to the absence of negative data in the pld – that is, data as to which constructions do not belong to the target language. For example, children are not generally informed that particular constructions that are ungrammatical are such, and when they themselves produce ungrammatical constructions in their speech they are not normally told that the construction in question is ungrammatical (Brown, 1973). And, in those rare cases where they are (for example, when the child has ambitious middle-class parents), the children seem impervious to the information and carry on to producing the ungrammatical construction regardless (Braine, 1976; Pinker, 1994). In the absence of such negative data, so the argument goes, the child learner runs the risk of overgeneralizing – that is, hypothesizing rules that generate all the sentences in the pld (so that the hypothesis fits her experience) but also generate further sentences that have not been encountered that do not belong to the target language. Without negative data, the child will have no opportunity to discover and correct any such mistakes.

4 Objections to Chomsky's nativism

Chomsky's claim that language development draws upon a substantial body of language-specific innate knowledge in the form of UG is, to put it mildly, highly controversial and has generated a lot of acrimonious debate. Thus, in her history of cognitive science, Margaret Boden (2006: 593) writes: 'I was astonished to discover the lack of mutual respect between the two camps in linguistics, Chomskyan and non-Chomskyan. I'd known there were unpleasant tensions of course, but I hadn't realized their degree.' In this section I will examine some of the most interesting and important objections to Chomsky's linguistic nativism and will strive to remain calm and polite in the process.

The existence of linguistic universals wouldn't prove that Chomsky's nativism was correct, as such universals could have some explanation other than the initial state of a biologically based language faculty. For example, one might argue that human languages are cultural inventions and all have basically the same form because they have been designed to deal with common problems that face all humans.[13] Alternatively, all languages might have a common origin in some early human language, so that they all bear the marks of this ancestor (Deutscher, 2005). However, although the existence of linguistic universals doesn't in itself vindicate Chomsky's position, if it turned out that there were no universals, then that would cause problems for him. In recent years a number of linguists and anthropologists studying non-European languages have argued that there are indeed no linguistic universals. Perhaps the most celebrated case is that of the Pirahã, a small and isolated Amazonian tribe who have been studied extensively by Daniel Everett (2005, 2012).

Everett (2012) thinks that languages are cultural tools that are designed by social groups to meet their needs. This opens up the possibility that languages vary quite dramatically on account of the divergent needs of the relevant social groups. Recall that Chomsky argued that syntactic rules are recursive. Indeed, in recent years he has placed even greater emphasis on recursion, portraying it as the central, universal feature of human language (Hauser et al., 2002; Fitch et al., 2005). Everett accepts that recursion is a feature of many languages because, within the cultural setting where those languages were developed, recursion helped communication. In particular, it enables complex messages to be communicated quickly and efficiently, avoiding redundancy and ambiguity. However, he claims, the language of the Pirahã lacks recursion. Therefore, rather than saying 'the smelly ferocious big black dog bit me', a Pirahã would say something like the following:

The dog bit me. The dog was black. The dog was big. The dog was ferocious. The dog was smelly.

And rather than saying 'Constantine thinks that Dan thinks that Steve thinks that Mark thinks that Paris is beautiful in the spring', a Pirahã would say:

Constantine spoke. Dan spoke. Steve spoke. Mark spoke. Paris is beautiful in the spring.[14]

The obvious problem with this absence of recursion is that there is a lot of redundancy and repetition in what is spoken. There is also potential ambiguity, as the hearer has to work out whether each distinct sentence refers to the same person, thing or event – for example, is it one and the same dog or a series of different dogs? However, argues Everett, within the context of Pirahã life, this isn't a problem for two reasons. First, the Pirahã are a close-knit tribe who know one another very well and use language only to talk about things and events that they have directly experienced or that have been experienced by someone close to them. Given this background, in the dog example they are likely to know that there has been only one dog attacking the speaker and that the speaker is likely to be talking about that one dog over the course of their monologue rather than a whole host of other dogs in addition to the biter. Second, the language of the Pirahã is an oral language (that is, it has no written form) and is typically spoken in a noisy context where there is a lot of outside interference with communication. The repetition helps the hearer to maintain focus, as each sentence brings them back to the topic of conversation. And, as large amounts of information are not packed into any single sentence, the costs of not hearing a sentence – due to, say, the howl of a monkey or the need to attend to a crying baby, or to throw a spear at a passing animal – is not the total destruction of the intended message.

I'm not convinced that the case of the Pirahã causes major problems for Chomsky and will argue that the language of the Pirahã might not appear to have recursion not because recursion is absent from their language but because they normally don't have the need to utilize it. In effect, their language has resources that are not usually drawn upon. Nevertheless, those resources are still a real feature of their language. A first step in this argument is to note the distinction that Chomsky (1965) makes between competence and performance: linguistic competence typically outstrips performance, so the latter is not the perfect guide to the nature of the former. This raises the possibility that the language of the Pirahã contains recursion, but for performance reasons the Pirahã don't usually draw upon it and so are akin to someone perfectly capable of running but who always walks because they never need to get anywhere quickly or because they don't want to break into a sweat. Such a bare possibility hardly counts as a strong argument; what is needed is some plausible explanation as to why the Pirahã don't utilize the recursive resources of their language. Perhaps Everett has provided us with a clue here in describing the Pirahã lifestyle as one of belonging to a close-knit community of multi-taskers living in a noisy environment.

When people know one another well and spend much of their lives in close proximity, they don't need to say much to communicate complex messages: their shared knowledge resolves problems of ambiguity and removes the need to use words to provide relevant background information.[15] For example, contrast how immediate family members talk when sat around the dinner table with how they talk when an unfamiliar guest is present. For the sake of successful communication with the guest, background information will have to be made explicit and care taken to clarify intended messages; the guest cannot be relied upon to have and utilize the kind of contextual information routinely employed by the family members. This will mean that more sentences are uttered and that those sentences are of greater complexity than when no guests are present. With respect to the Pirahã, what this suggests is that, as they are a small close-knit community that spends much of their time together, they may well not need to rely upon the complex sentences that their language contains in order to communicate what they want to communicate.

In certain contexts all of us simplify the way we speak in order to facilitate communication. One of the contexts in which we do this is when we are aware that outside factors are likely to interfere with our attempts to communicate or when we, or our audience, are multi-tasking and so not able to process complex sentences. Multi-tasking is very difficult for humans, as we find it very hard to attend to, or concentrate on, more than one thing at once (Chabris and Simons, 2010). For example, when driving, those at the wheel are often forced to simplify their speech if talking to a passenger in order to avoid crashing. Interestingly, evidence suggests that passengers are sensitive to the difficulties facing the driver and so tend to simplify their own speech, and this explains why talking to a passenger while driving is much less dangerous than talking on a mobile phone while driving. As Chabris and Simons put it:

> When you converse with the other people in your car, they are aware of the environment you are in. Consequently, if you enter a challenging driving situation and stop speaking, your passengers will quickly deduce the reason for your silence. There's no social demand for you to keep speaking because the driving context adjusts the expectations of everyone in the car about social interaction. (2010: 26)

Returning to the Pirahã, my suggestion is that, because they are usually multi-tasking when they speak and are likely to have their conversations interrupted by external events, they are pressurized into keeping their sentences artificially simple. In short, then, I'm suggesting that we shouldn't read too much into the typical simplicity of the sentences produced by the Pirahã, as contextual factors may well be pushing them away from utilizing the full recursive resources of their language on a regular basis.

Evans and Levinson (2009) provide another important critique of the claim that there are language universals. They argue that the approach within the Chomskyan generative tradition has focused upon English and its European relations, ignoring languages from other parts of the world. When one takes into account those languages, they continue, both the idea that there are language universals and UG become unsustainable. As they put it:

> languages differ so fundamentally from one another at every level of description (sound, grammar, lexicon, meaning) that it is very hard to find any single structural property they share. The claims of Universal Grammar, we argue here, are either empirically false, unfalsifiable, or misleading in that they refer to tendencies rather than strict universals. (2009: 429)

As we have already seen, putative exceptions to postulated universals need not be as convincing as they first seem. This point can be bolstered by noting an important methodological feature of Chomsky's approach. The identification of a particular universal[16] is not typically based upon the comparison of many distinct languages that have been independently analysed. Rather, what is driving the generative approach is the need to explain language acquisition. Hence, if an account of a particular language makes it difficult to see how its speakers could have acquired it, then that counts heavily against that account even if it constitutes an elegant description of the language. UG was postulated as a response to the poverty of the stimulus considerations in order to explain how children could acquire languages with abstract syntactic rules despite the limited nature of their linguistic experiences during the language acquisition period. The idea, then, is that UG must exist to make language acquisition possible. This leaves it an open empirical question as to what form UG takes, and so any suggestion within the principles and parameters framework is simply a hypothesis that might turn out to be true or false. To be viable, any such hypothesis must facilitate the explanation of the acquisition of any given human language in the circumstances in which it is normally acquired, and this is why cross-linguistic analysis plays a prominent role in the generative tradition, whatever Evans and Levinson say.

Now suppose that we accept Evans and Levinson's assertion that all the claims that have been made by Chomskyans so far as to what is universal (for example, those listed by Pinker and Bloom, 1990) have been falsified by the discovery of some language lacking the feature in question. This doesn't compel one to reject the idea that UG exists, as one could take the line that the task of uncovering the nature of UG was bound to be a hard one, that we have yet to crack it, but that the search must continue. Whether or not one takes this latter option will of course depend upon how convinced one is by the poverty of the stimulus considerations. For their part, Evans and Levinson are not concerned primarily with explaining language acquisition, but they do concede that their perspective makes doing this even more of a challenge, for two reasons:

> First, because the extraordinary structural variation sketched in this article presents a far greater range of problems for the child to solve than we were aware of fifty years ago; and second, because the child can bring practically no specific hypotheses, of the UG variety, to the task. (2009: 445)

For me, the upshot of all this is that the crucial questions are those of how powerful the poverty of the stimulus argument is and whether there is any currently available theory that plausibly explains how children learn their first language. The remainder of the chapter will focus on these questions.

Not everybody has been convinced by the poverty of the stimulus argument. A number of critics – for example, Pullum and Scholz (2002), Scholz and Pullum (2006), Cowie (1999) and Sampson (2005) – have argued that the pld is not as limited as Chomsky has made out. With respect to the rule for forming questions considered above, they argue that complex questions such as 'is the man who is happy singing?' are present in the pld. They don't make this claim on the basis of examining what is directly said to children. Rather, they do it on the basis of corpus analysis, where the corpuses analysed include the *Wall Street Journal* corpus and Oscar Wilde's play *The Importance of Being Earnest*. How seriously we should take these considerations depends on how representative the cited corpuses are. Pullum and Scholz take them to be representative and go so far as to assert that 'a construction inaccessible to infants during the language acquisition process must be rare enough that it will be almost entirely absent from the corpora of text quite generally' (2002: 21). However, they produce no independent argument for this claim. Nevertheless, they do argue that we should realize that children are not just exposed to 'baby talk' addressed directly to them but hear an awful lot of fully fledged adult language when they watch TV, overhear adult conversations, listen to pop songs, and so on.

Even if this last claim about what sentences children are exposed to is true, there are two powerful responses open to the supporter of Chomsky. First, one aspect of the impoverished nature of the pld is that it contains many ungrammatical constructions that potentially mislead children if they take them into account when trying to learn their first language. So, if children are ranging as widely as Pullum and Scholz (2002) suggest, they will be attempting to learn on the basis of seriously compromised data, making it difficult to see how they could learn on the basis of the pld.

Second, it is not merely a question of what children hear but what they take into account in attempting to learn language. The worry is that, if a child attempts to take into account everything that they hear, they will be overwhelmed by data. For example, encountering complex constructions of the kind presented by Sampson (2005), Pullum and Scholz (2002) and Cowie (1999) will be of little help to the child if she cannot remember them for anything more than a fleeting moment. She needs to remember such constructions accurately for enough time to work out their implications given what she has previously encountered. This will place a heavy burden on the child that she may well not be able to meet. As Crain and Pietroski (2001: 147) point out, 'Adults can recall (at best) the gist of the immediately preceding word string, not its phonological or syntactic details. Surely children cannot be expected to remember more than adults do.'

What this last point hints at is that, when considering the poverty of the stimulus argument, there is a tendency to interpret it as saying merely that a child is exposed to a very meagre range of sentences when developing language. However, the argument should be understood as pointing out that, no matter what belongs to the pld, for that input to be helpful to the child it needs to be packaged and presented in an appropriate way. Such packaging would provide the child with the scaffolding or the guidance to make her way through a body of data that is potentially overwhelming. Hence, corpus analysis on its own does little to undermine the poverty of the stimulus argument.

Another response to the poverty of the stimulus argument directly tackles the claim that the pld is full of ungrammatical structures. According to this, parents and carers typically talk to children in a distinctive way: they produce clear and short sentences; they deliver these sentences with a sing-song intonation; they restrict their talk to things and events that are readily perceivable in the child's immediate environment; and they use overt gestures to indicate what they are talking about. For their part, young children are disposed to focus on such language, preferring it to other forms of speech. This way of speaking to children is known as Motherese or Child Directed Speech. The developmental psychologists Gopnik, Meltzoff and Kuhl (1999) go so far as to say that children have evolved to prefer listening to Motherese to other forms of language and that all people (including older siblings who are themselves children) have evolved to speak automatically in Motherese to young children. The upshot of this is that children have a very clean body of data to help them learn language (Snow, 1977; Newport et al., 1977); in effect, Motherese constitutes a mechanism that protects children from all the ungrammatical structures said within earshot when adults talk to other adults.

There is an obvious objection to this appeal to Motherese. If children do focus on Motherese, they may well be getting clean data. However, it will be limited in its complexity, giving rise to the worry that they will alight on rules that fit the Motherese pld but don't work for language in general.

The upshot of these considerations is that a challenge faces opponents of the poverty of the stimulus argument. This is the challenge of explaining how a child steers a course between, on the one hand, being overwhelmed by a mass of data, much of which is 'dirty' and so potentially misleading, and, on the other hand, having to rely on a 'clean' body of data that is unrepresentative of the target language.

I will now turn to the logical problem of language acquisition. Fiona Cowie (1999) has developed an important objection to this version of the poverty of the stimulus argument that is echoed by Michael Tomasello's (2003) attempt to explain how children avoid overgeneralizing. Cowie argues that, even if negative data[17] is not available to the child learning language, it does not follow that children do not have access to negative evidence that can serve to disconfirm their faulty overgeneralizations. This is because there is a distinction between negative evidence and negative data. As Cowie puts it:

> [P]ositive data can constitute *both* positive *and* negative evidence for a theory – and similarly for negative data. The positive datum that *a* is G can be positive evidence for (i.e. confirm) a theory T if, for instance, T predicts that *a* is in fact G; conversely, if T predicts that *a* is not G, then that same positive datum will be negative evidence for (i.e. disconfirm) T. Similarly, the negative datum that *b* is not G can be positive evidence for T (if T predicts that fact) or it can be negative evidence for T (if T predicts the contrary). (1999: 222)

In order to illustrate and motivate her position, Cowie presents an example involving the refutation of the hypothesis that all intransitive verbs can be used as causatives.[18] This implies that 'I falled the cup off the table' is grammatical. Cowie describes a child, Edna, floating this hypothesis. She subsequently sees her father knock a cup of coffee off a table and hears him say 'I caused the cup to fall off the table', providing her with the positive datum that that construction is grammatical. But this positive datum provides evidence against the hypothesis in question, as that hypothesis would suggest that Edna's father would say 'I falled the cup off the table'. In reasoning thus, Edna seems to be assuming that, if a verb is used in a sentence that serves a particular communicative function, then alternative possible forms that serve the same function will be ungrammatical. Tomasello (2003) explicitly attributes this assumption to children and calls the phenomenon of acting on it 'pre-emption': what a child hears pre-empts her from making an illegitimate generalization (in this case, that all intransitive verbs can be used as a causative). Tomasello argues that something similar goes on in the case of learning the meaning of words: children assume that most words have distinct meanings, so that when they hear a new word they do not attribute to it the same meaning as any other words they know. According to Tomasello, pre-emption is particularly important because it grounds adult correction. For example, when an adult rephrases a child's sentence, the child assumes the adult is expressing the meaning correctly, so that it pre-empts what she herself said. In Cowie's terminology, this would be a case of a positive datum (that the sentence produced by the adult is grammatical) providing negative evidence (that the sentence produced by the child is ungrammatical).

I'm far from convinced by this response to the logical problem of language acquisition. First of all, consider linguistic creativity, which implies that children are frequently producing and encountering perfectly good sentences that they have never produced or encountered before and that the stock of such sentences is limitless. Wouldn't a child who was alive to this feature of language expect that, with respect to any given situation, there are many different legitimate ways of describing it, many involving unfamiliar sentences? If this were the case, then a child would hardly be likely to conclude that, because she had not heard a particular sentence or sentence form, that sentence or form was ungrammatical.

Now consider the assumption attributed to children by Cowie and Tomasello. To recapitulate, this is the assumption that, if a verb is used in a sentence that serves a particular communicative function, then alternative possible forms that serve the same function will be ungrammatical. I find it difficult to see why a child would make that assumption, as it is inconsistent with her experiences; there are often many different ways of communicating a given message. Suppose Edna builds a snowman on a winter's day. Later in the day the sun comes out and the snowman quickly melts. One could describe this event with any of the following sentences featuring the intransitive verb 'melt': 'the sun melted the snowman', 'the sun caused the snowman to melt', 'the sun made the snowman melt', 'the snowman was melted by the sun', 'the snowman was caused to melt by the sun', 'the snowman was made to melt by the sun', 'the snowman's melting was because of the sun', and so on. Given this, why wouldn't Edna in Cowie's example assume that 'I caused the cup to fall off the table' was just one option chosen by her father from many equally legitimate alternatives, one of which is 'I falled the cup off the table'? If, however, what her father says does pre-empt the latter sentence, then why doesn't hearing her father say 'the sun melted the snowman' also pre-empt 'the sun made the snowman melt'?

Another point that needs to be borne in mind is that which sentence an individual produces in a given situation is an interaction effect where grammaticality is only one of the factors involved. For example, a given individual or group might accept that a particular sentence is grammatical but resist uttering it for a whole range of factors: they might view it as inelegant, offensive, old-fashioned, ambiguous, pretentious, and so on. So a child had better not regard the mere non-occurrence of a construction as evidence for its being ungrammatical or she will be led astray in the course of language development. Similarly, if an adult rephrases what the child says, she needs to be wary of concluding that her utterance was ungrammatical; it may well just be that the adult prefers an alternative way of putting things for extra-grammatical reasons. What this point brings out is that, even if we ignore the many grammatical mistakes that people make in speaking, the data provided by the pld is heavily muddied. In order to negotiate her way successfully through it, a child learner is going to need to have and bring to bear a knowledge of a whole range of extra-grammatical factors that influence what people do and do not say. In sum, the child learner's task is not as straightforward as Cowie and Tomasello suggest.

5 Tomasello's usage-based account of language acquisition

Over the course of two prominent books – namely, *Constructing a Language* and *Origins of Human Communication* – Michael Tomasello has developed an approach to language that constitutes a hugely important challenge to that of Chomsky and his followers.[19] Tomasello calls his approach 'a usage-based theory', and in doing so he is making more than a nod to Wittgenstein (1953) with his emphasis on how we use language in the course of our variegated everyday lives and activities. In this section I will provide an account of Tomasello's perspective and address the question of how well it fares as an alternative to linguistic nativism.

Tomasello doesn't regard language as a biological entity or as based on evolved cognitive capacities that are specific to language. Thus, he has no truck with the postulation of Universal Grammar or a language faculty. Nevertheless, based upon his extensive research on the cognitive capacities of our great-ape relatives, he follows Chomsky in accepting that language is specific to humans and that chimpanzees, gorillas, and the like, are not capable of developing genuine linguistic capacities. What sets humans apart from all other species relates to their pattern-finding and mind-reading capacities, which are aspects of our evolved, biological endowment. It is these capacities that have been brought to bear by communities of humans in creating languages and by children (typically in conjunction with their adult carers) in learning the language of their community.

What exactly are these pattern-finding and mind-reading skills that underpin language? Consider pattern-finding first, which I will explain via an example. Suppose an individual frequently suffers from an allergy-based skin rash. She assumes that the occurrence of this rash is not random but exhibits some form of regularity or pattern. She seeks to uncover this regularity or pattern in the hope that it will reveal the underlying causes of her allergic reactions, so enabling her to manage her condition. With this aim in mind, she keeps a detailed diary that for each day specifies the timing and severity of any skin rash. In addition, for each day she documents everything she eats, her activities, psychological state, cleansing practices, and so on. After three months she turns to processing this data, and this involves a good deal of statistical analysis. For example, she works out what proportion of the days when she was stressed she developed a rash as against the proportion when she was relaxed. Similarly, she works out what proportion of days when she ate a particular food item (say, wheat, rice, milk or strawberries) she developed the rash as against the proportion when she didn't eat the relevant item. And so on, for all the factors she records. What this statistical analysis reveals is that she experiences the rash in a severe form when and only when she is either very stressed or has eaten strawberries. In other words, the pattern that she uncovers on the basis of examining the data relating to her lifestyle and experiences is that severe rashes are strongly correlated with both stress and strawberry consumption but not with any other factors relating to diet, psychological state or behaviour.

Another familiar example of pattern recognition includes that underlying the continuation of the number series 1, 2, 4, 8, 16 . . . with the number 32 (in this case, the identified pattern is that each number is twice the size of its immediate predecessor). This can also be regarded as a case of statistical learning, as identifying the relevant pattern involves uncovering a statistical regularity in the input data (all of the numbers in the series are twice the size of their immediate predecessor).[20]

By 'mind-reading', I mean the everyday activity of working out the mental state of another person in order to explain or predict their behaviour or some aspect of their mind. For example, suppose I see you queuing at a café and conclude that you want a cup of coffee and intend to buy one from the café, then I'm mind-reading.[21]

Tomasello thinks that neither pattern-finding nor mind-reading is specific to humans, but he does think that our mind-reading skills have a dimension that even those of our closest relatives don't have and that this relates to attributing higher-order mental states. Consider my belief that Paris is beautiful in the spring or my intention to make a cup of coffee in half an hour's time. These are first-order mental states. Now suppose that I believe that my colleague Steve believes that Paris is beautiful in the spring or I intend to get you to want a cup of coffee. Then my mental states (my belief in the first case and my intention in the second) are higher-order (rather than first-order) mental states. In general, a higher-order mental state is a mental state the having of which involves attributing a mental state to someone (either oneself or another person).

Tomasello follows Grice (1957, 1975) in holding that higher-order mental states are central to linguistic communication. To see what this comes to, consider an example. I say to my colleague Steve, 'Can you get me a cappuccino from the café?' My ultimate intention in saying this is to get Steve to get me a cappuccino from the café. But the crucial point is that I aim to do this by getting him to token a higher-order mental state that involves his being aware of my intentions. So, I intend to get Steve to believe that, in saying what I did, I intended him to believe that I wanted him to get me a cappuccino from the café. If all goes well, he will indeed token such a seriously higher-order mental state and act on it.

According to Tomasello, the situation in the above example is typical of linguistic communication; that is, by its very nature, linguistic communication involves the participants having higher-order mental states relating to the higher-order mental states of one another. This has some serious implications. First, when it comes to our mind-reading skills we must be capable of recursivity – that is, of embedding any given mental state in a more complex mental state. Second, we must have highly developed mind-reading skills in order to engage in successful linguistic communication. In the above example, to understand me, Steve needs to be able to work out my communicative intentions when I speak and so needs to be correct in his beliefs about what I intend. For my part, I have to be able to know what to do to get him to form a belief that correctly identifies my communicative intentions.

Tomasello is keen to stress that not all communication is linguistic. Suppose a drinker in a bar catches the bartender's eye and points at his empty glass. Understanding his communicative intentions, the bartender fills the drinker's glass. Again, suppose a diner in a restaurant mimics twisting a pepper grinder in view of the waiter, who responds by bringing a pepper grinder to the table.[22] Actions such as these happen all the time but don't have a fixed meaning. For example, pointing to an empty glass in a bar could mean 'My glass is empty. Can you fill it?' But when done by a recovering alcoholic it could mean 'Look, I've not had anything to drink'. Thus, the meaning of such gestures is context relative. What is central to context is what Tomasello calls 'common ground' (2008: ch. 3). This is something that is shared between the communicators and something that they know they share and involves such things as being involved in a particular activity, having a common body of knowledge, having a shared history, and so on. Thus, the drinker and bartender both have a knowledge of how bars work (for example, that customers are there to drink, that bartenders are there to serve them), the customer knows that the bartender is a bartender, and the bartender knows the customer is a customer, and so on. The recovering alcoholic is communicating with his Alcoholic Anonymous mentor who knows him to be a recovering alcoholic, and the recovering alcoholic knows the mentor is there to help him and ensure he doesn't drink, and so on. Without such common ground we wouldn't be able to make any sense of such gesturing and pantomiming. Once again, higher-order mental states are prominent: not only does communication require working out communicative intentions, as in the linguistic case, but in working these out we must attribute to our fellows bodies of understanding along with a knowledge that we both know that we have this mutual knowledge.

For Tomasello, linguistic communication is a lot like these examples of non-linguistic communication in that it involves the attribution of higher-order mental states and the existence of a common ground. One key difference, though, relates to the centrality of conventions in the linguistic case. Echoing many philosophers,[23] Tomasello thinks that conventions are central to language: there are conventions governing the use of language. To pick a simple example, it is conventional in English to use the word 'glass' to refer to a certain kind of drinking vessel. Speakers of a language have knowledge of these conventions and know that their fellows also have this knowledge. The existence and knowledge of linguistic conventions form part of the common ground that makes linguistic communication possible. Without such conventions to bind us together, it would be much harder for a speaker to find a way to express her communicative intentions in such a way that they are likely to be understood and much harder for the hearer to work out the speaker's communicative intentions. Just think how difficult it is communicating with someone with whom you don't share a language.

If language aids communication then how do we acquire language? Tomasello thinks that children learn language – there is no UG or language faculty – and he develops a detailed account of how that learning process runs. Many attempts to resist Chomskyan nativism are undermined by a failure to provide any realistic account of what an individual learns when she learns a language. Typically, they highlight some relatively simple aspect of language and construct an account of how that aspect could be learned on the basis of data that is readily perceivable and available in the child's environment. They then postulate that all other aspects of language are learned in the same way without giving any detailed consideration to these aspects of language.[24] This charge, however, cannot be directed at Tomasello, as he has a detailed conception of what is involved in knowing a language that draws upon an approach in linguistics known as construction grammar.[25]

The central concept of construction grammar is, not surprisingly, that of a construction. A construction is a 'symbolic unit with meaning' (Tomasello, 2003: 160). Thus, any particular concrete word or sentence (such as 'dog', 'chased', 'the dog chased the postman', and so on) is a construction. But many constructions are abstract, as concrete sentences exemplify more abstract forms. For example, the concrete construction 'the dog chased the postman' exemplifies the following more abstract constructions:

- X chased the postman.
- X chased Y.
- TRANSITIVE-SUBJECT TRANSITIVE-VERBed TRANSITIVE-OBJECT.

Each construction in this list is more abstract than the one that preceded it. The first two are examples of item-based constructions, as they feature at least one concrete word or lexical item. The third is an abstract utterance-level construction. For Tomasello, an individual's knowledge of their language is constituted by an inventory of constructions ranging from the concrete to the highly abstract. Thus, when an individual utters a particular sentence in order to communicate a specific message (or understands the communicative intentions of someone else when they uttered a particular sentence), they draw upon constructions represented in their mind. If an abstract construction is involved, then particular concrete words are placed into the slots that figure in that abstract construction.

The more abstract constructions might look like the kind of rules or principles Chomsky postulates (such as 'the head of a phrase comes before its complement'). However, this appearance is misleading. Tomasello describes Chomsky's approach as a 'formalist' one, as it portrays the core of language as a system of formal syntactic rules that themselves have no meaning. Thus, the meaning of a sentence comes from the words that are put together by using the syntactic rules. Constructions, on the other hand, do have meaning. Here is how Tomasello makes the point:

> [C]onstructions are themselves meaningful symbols – since they are nothing other than the patterns in which meaningful linguistic symbols are used to communicate. Thus the pattern *X VERBed Y the Z* is a construction

of English that signifies some transfer of possession (either literal or meta-phorical); the pattern *the X* signifies a 'thing' . . .. There are no linguistic entities – lexical or syntactic – that are not symbolic in this way; all have communicative significance because they all derive directly from language use. (2003: 99)

As knowing a language involves having an inventory of constructions represented in one's mind, learning a language involves learning such an inventory. Tomasello offers an account of this process that portrays it as a gradual and piecemeal one where there is no clean break between syntax and semantics. It is in this context that pattern recognition and mind-reading loom large, as we will now see.

An infant will come to the language learning process with certain perceptual and cognitive capacities in place:

1   the possession of certain basic concepts and the ability to perceive a viewed scene in terms of those concepts. This skill would be involved when, for example, a child, seeing Bill throw a ball to Jim, parsed the scene into three distinct objects – two people and a ball – and saw one of the people as acting on the ball in such a way that it moved from their possession to that of the other person;
2   pattern-recognition mechanisms that enable her to construct representations that pick out abstract similarities between distinct items that differ at the concrete level;
3   mind-reading skills that include an ability to discern higher-order mental states of the kind that are involved when someone does something with a communicative intention.

From her earliest days a child will be involved in certain oft-repeated, routinized activities with her carer – for example, being fed, being dressed, playing with a toy, and so on. In this context, the child and carer will establish a joint attentional frame. That is, there will be aspects of the scene that both perceive they will not only both attend to but be aware of their mutual attention. For example, if the pair are playing with a ball, they will both attend to the ball, know that the other is attending to the ball, know that the other knows this, and so on. Here, then, we have an example of the exercise of mind-reading skills involving higher-order mental states. Suppose that in this situation the adult says 'ball gone' immediately after hiding the ball behind her back. The child's task is to work out the adult's communicative intentions in saying this and so represent the meaning of that sentence so as to form an item of linguistic knowledge. The fact that the pair have been jointly attending to the ball in the context of a regular and familiar activity and that they mutually know that the ball has recently disappeared, and that this event jointly amuses both of them, gives the child a massive clue as to the adult's communicative intentions. That clue steers her away from concluding that the adult is talking about any number of things that can be readily perceived, such as the rug they are sitting on, the cat that is passively watching the scene, the TV in the corner of the room, and so on. Hence, the child concludes that the adult intends to say that the ball has disappeared.

Now suppose that the watching cat stirs noisily, attracting the attention of the adult. She turns to look at the cat, turns back to the child and then looks at the cat again as it slinks out of the room, having drawn the child's attention to the cat. The adult then says 'cat gone'. The clues here will enable the child to appreciate that the adult intends to say that the cat has disappeared, and thus she adds to the stock of linguistic representations stored in her mind. At this stage the child is like someone learning a foreign language from a phrasebook. That is, she knows the meaning of various sentences on a one-by-one basis, but she doesn't appreciate the contributions made to their meaning by their components and so cannot construct any new sentences. In other words, she lacks linguistic creativity. However, this situation changes as the child comes to appreciate that sentences with similar meanings contain recurring sounds. For example, both 'ball gone' and 'cat gone' talk about something disappearing and contain the sound 'gone'. This enables the child to work out that 'gone' relates to disappearance. In a similar manner she works out that 'cat' refers to cats and that 'ball' refers to balls. But she is not only able to work out the meaning of individual words by reflecting upon the similarities between different sentences that are initially understood in terms of their overall meaning. For, in addition, she is able to recognize that there are structural patterns exemplified by distinct sentences containing a common word. Thus, for example, both 'cat gone' and 'ball gone' have the form 'X gone' where 'X' is a linguistic item that refers to an object. Realizing this, the child stores the construction 'X gone', representing it as meaning that thing X has disappeared. This construction can then be utilized to enable the child to understand and construct new sentences of that form that she has never encountered before. What is particularly noteworthy about this process is that identifying constructions relating to particular words and those relating to larger structures in which those words appear goes hand in hand and is mutually supportive.

At this stage, then, the child has placed a number of concrete words and sentences and some more abstract item-based constructions in her inventory of constructions. To gain full mastery of a language she will have to add a rich stock of abstract utterance-level constructions. Tomasello (2003: ch. 5) provides an extensive list of these, including the following:

> Simple transitive, exemplified by 'the boy kicked the ball':
> TRANS-SUBJ TRANS-VERB TRANS-OBJ

Simple intransitive, exemplified by 'the vase broke':
INTR-SUB INTR-VERB

Ditransitive, exemplified by 'he gave me a book':
DITR-SUB DITR-VERB DITR-R DITR-O

Dative, exemplified by 'I sent a package to Minneapolis':
DA-SUBJ DA-VERB DA-O to DA-L

'Content as object' locative, exemplified by 'she loaded hay
onto the wagon':
LOC-SUBJ LOC-VERB to/on/with/etc. LOC

'Location as object' locative, exemplified by 'she loaded the wagon
with hay':
LOC-SUBJ LOC-VERB with/from/etc. LOC-OBJ

Resultative, exemplified by 'he wiped the table clean':
RES-SUBJ RES-VERB RES-OBJ RES-ADJ

Passive, exemplified by 'Spot got hit by a car' and 'Bill was shot by John':
PASS-SUBJ *get* VERB-*ed by* PASS-OBJ
PASS-SUBJ *be* VERB-*ed by* PASS-OBJ

In order to learn these abstract utterance-level constructions, argues
Tomasello, the child once again uses her pattern-recognition skills. This
involves finding analogies: 'children begin to form abstract utterance level
constructions by creating analogies among utterances emanating from
different item-based constructions' (2003: 163–4). Doing this involves
aligning distinct elements of the sentences considered on the basis of their
functional role in the sentence to which they belong. So, for example, suppose a child compares 'the dog chased the postman' and 'Bill kicked the
ball'. She can then appreciate that 'the dog' and 'Bill' play the same role in
their respective sentences, as do 'chased' and 'kicked' and 'the postman'
and 'the ball', respectively. On the basis of considering a whole series of
analogous sentences, she generalizes to arrive at the construction TRANS-
SUBJ TRANS-VERB TRANS-OBJ. Armed with this, she can now produce
and understand previously unencountered sentences by fitting relevant
words into this construction.

Abstract utterance-level constructions invoke abstract syntactic roles
such as SUBJECT and OBJECT as well as abstract grammatical categories such as NOUN and VERB. The question arises, then, of how such
notions are formed. Once again, pattern-finding is involved, but the

precise pattern-finding processes reflect a key difference between syntactic roles and grammatical categories. The grammatical category to which an item belongs is fixed and is the same whatever sentence it appears in. However, a particular item might occupy different syntactic roles in different sentences. For example, 'the dog' is the subject in 'the dog chased the postman', the direct object in 'the postman chased the dog', and the indirect object in 'Mary gave a bone to the dog'. Thus, syntactic roles can't be equated with form.

The pattern-finding process involved in forming the syntactic roles such as SUBJECT and OBJECT is, once again, that of analogy, as the child appreciates that differing items in distinct constructions play a common role, as do 'the dog' and 'the postman' in the respective sentences 'the dog bit the postman' and 'the postman chased the dog'. With respect to forming such grammatical categories as NOUN and VERB, a different pattern-finding process is involved, namely, functionally based distributional analysis. Tomasello describes this process well when he writes:

> [C]hildren form paradigmatic categories of linguistic items – either words or phrases – that play similar roles in the utterances they hear around them. Thus, *pencil* and *pen* occur in many of the same linguistic contexts in utterances – they do many of the same kinds of things in combining with articles to make reference to an object, in indicating subjects and objects as syntactic roles, and so on – and so a language user will come to form a category containing these and similarly behaving words. (2003: 170)

In sum, then, according to Tomasello, learning a language involves storing constructions, many of which are highly abstract. This is a gradual and piecemeal process that involves (i) the utilization of mind-reading skills in working out the communicative intentions of a speaker in the context of shared activities where a common attentional frame has been established; and (ii) the application of pattern-recognition skills to constructions that have already been stored.

I now turn to the question of whether Tomasello offers a more plausible account of language acquisition than Chomsky. A first point relating to this question has to do with the mind-reading skills of young children. Tomasello argues that language learning begins in earnest around the age of one year because at this stage children become able to enter into rabidly higher-order mental states and attribute such states to others, something that chimps are unable to do. Rabidly higher-order mental states are involved in establishing joint attentional frames. For example, if an adult and a child are jointly attending to a ball, both must attend to the ball, know that the other person is attending to the ball, and know that the other person knows that they are both attending to the ball. Moreover, when something is said by the adult in this context and the child understands them, the child must recognize the communicative intentions of the adult. As we have seen, Tomasello conceives of communicative intentions in Gricean terms, so that when the child understands the adult's utterance 'ball gone' she appreciates that the adult intends to cause the child to believe that the adult intends, in saying 'ball gone', to cause the child to believe that the ball has disappeared.

One might reasonably ask whether children as young as one have such sophisticated mind-reading skills. There is considerable evidence suggesting that children have mind-reading capacities from a very early age. Very young babies distinguish between people and inanimate objects; for example, if an inanimate object stops moving they lose interest in it, but if the face of a person they have been interacting with falls still they get very upset (Tronick et al., 1978). By age one, babies are sensitive to the expression of emotion; for example, if a baby is crawling towards an area and their parent makes a facial expression of disgust or horror, the baby will tend to avoid that area (Campos and Stenberg, 1981). This shows an understanding of the emotional state of their parent. However, many psychologists think that young children do not have a fully fledged ability to mind-read until at least age four, as is evidenced by a famous psychological experiment. This is called the false belief test or, alternatively, the Sally-Ann test.[26] Here is how a classic version of this test goes. A child watches a puppet show featuring two puppets, one named Sally and the other named Ann. They are in a room that contains a basket and a box. Sally has a marble that she places in the basket. She then leaves the room. Ann takes the marble out of the basket and hides it in the box. The watching child is asked where Sally will look for the marble when she returns to the room. The correct answer is that she will look in the basket, as she will act on the false belief that the marble is in the basket where she left it (recall that Sally was out of the room when Ann moved the marble and so does not know of the shift). Whether or not the child gives the correct answer is sensitive to their age. Children younger than four generally predict that Sally will look in the box, whereas those four and older generally give the correct answer. What this suggests is that, before the age of four, children fail to appreciate that other people can have beliefs that differ from their own beliefs, that other people can have beliefs that are false by their lights. The under-fours subjected to this test just assume that Sally believes what they themselves believe. After four, children appreciate that another person can have a quite different take on the world and so can have beliefs that are, by their lights, false. Consequently, they have much more effective mind-reading skills. Therefore, our mind-reading skills undergo a process of development in early childhood.[27]

What this might be taken to suggest is that children are unlikely to be as good at mind-reading as Tomasello requires until they are about four. But this is way beyond the age that Tomasello suggests children are learning language by attributing rabidly higher-order mental states. However, in recent years the water has been muddied somewhat, as some psychologists have argued that children are capable of attributing false beliefs much earlier than has traditionally been assumed. The key figure here is Renée Baillargeon, who argues that the failure of young children to pass the traditional version of the test is due to the complex language in which they are instructed as to what they have to do. In a nutshell, they don't understand what the experimenter is asking them to do. However, she has devised simpler versions of the false-belief test that children are able to pass much earlier (as early as one year) (Baillargeon et al., 2010).

Suppose that we accept that children do have the mind-reading skills Tomasello attributes to them and that it is in principle possible for them to learn a language in the manner he describes. This wouldn't in itself vindicate his theory, as it would need to be established that children actually learn language in the way he suggests. This point isn't merely academic, as there are grounds for being sceptical that children generally receive the kind of structured input that his theory requires.

Tomasello portrays adults as working very hard to help the children in their care learn language. They have to establish and maintain joint attentional frames and in that context produce sentences of the appropriate level of complexity. Anyone who has had much contact with young children (especially in the age range one to four) will recognize that such children often have problems maintaining focus and frequently 'zone out' when one is trying to engage them. They don't expect to understand everything said to them and typically don't work hard to understand things said to them that they don't initially understand; in this respect they are very different from adults. Thus, maintaining a joint attentional frame requires great skill on the part of the adult: they have to keep the child's interest and say things that are not so banal that the child becomes bored or so hard to understand that the child zones out.

Recall that Tomasello argues that language learning is an incremental process: a child learns language gradually, something that involves building and storing increasingly complex constructions. Now there are examples of systematic incremental childhood learning that occur within joint attentional frames where adults work hard to facilitate childhood learning. One of these relates to mathematics education in primary schools. Here is a sketch of how that process goes in the United Kingdom. Mathematics is part of the national curriculum, so firm targets are specified by central government as to where a child should be at any given stage in their education. Teachers teach to the national curriculum and are trained to do so. This involves regular measuring and assessment to determine a child's capacities at any given point in time and to uncover any problems and failures to hit targets. The teacher's approach to mathematics education is very carefully designed. Different mathematical skills are identified and taught in a very specific order, with each stage building on what has gone before. For example, mathematics education will begin with counting and then move on to addition involving numbers between one and ten (often illustrated with concrete objects, so utilizing basic counting skills). Next comes subtraction, then multiplication and finally division. After that, more demanding topics are covered. Consider the case of learning times tables. This is done in a specific order (1, 2, 10, 5, 4, 3, 8, 6, 9, 7); so as to ensure that the child masters each times table before moving on to the next, the teacher gets the child to recite each one over and over again at regular intervals and take weekly tests. By this means most children become competent at mathematics by the time they leave primary school. Even so, some children do not develop the skills expected of them even when they have the relevant cognitive capacities. This can be due to factors such as absence from school through illness, incompetent teaching, or disengagement with school life. This indicates that the learning process is a fragile one that can easily be knocked off course.

The existence of such a clear-cut case of incremental learning within the context of joint attentional frames formed with an adult might be taken to support Tomasello. If it frequently happens in the case of mathematics, then why can't it happen in the case of language, so the reasoning might go. However, there is a worry that the case of mathematics constitutes an implausible model for language learning. To see this, consider the following.

Children begin school at age four in the UK and so their language skills are well in place before they receive any formal teaching in school. This means that their parents and carers need to have played the equivalent role to that of a teacher in the case of mathematics. Are most parents and carers capable of playing this role? By anyone's reckoning, the knowledge and skills involved in speaking a language are far more complex and demanding than those involved in basic mathematical competence. I doubt that any more than a handful of parents and carers have any detailed idea of how to break down mature language skills into simple components and identify which are the easiest to master. Therefore, few parents or carers will be in a position to design and execute the equivalent of curriculum. They will have no idea what to be working on at any given time, how much practice needs to be done before moving on to the next task, how often to return to a previously mastered skill so that it isn't forgotten, and so on.

Another problem relates to a further disanalogy with the mathematics case. Most primary school children don't practise mathematics with one

another and so they are not attempting to learn with, or from, their contemporaries. This is clearly a good thing: think what could go wrong if a
five-year-old were trying to learn mathematics from another five-year-old.
Yet much of the language-using behaviour of children is engaged in with
their contemporaries; that is where the bulk of a child's engagement in
joint attentional frames takes place. But what this means is that they are
hardly likely to get the high-quality input that Tomasello's theory requires.
On the contrary, their input is going to be full of errors and lacking the
structure and organization it needs to best facilitate the learning process.

To cement my scepticism, let's return to the initial example of the child
learning the 'X gone' construction. I accept that this is a perfectly plausible
scenario and that it does explain how that construction could be learned.
But it's important to notice two features of the scenario. First, the adult
is working very hard to make her communicative intentions clear to the
child; the two are engaged in a familiar activity that both of them find
engaging, and the adult has cues at hand that she can readily manipulate.
Suppose that the child indicated she hadn't understood the first utterance
of 'ball gone'. Then the adult could be relied upon to bring the ball back
into view and then remove it and say, once again, 'ball gone', so giving
the child a second chance at picking up the message. Second, if the adult
left it at that, the child would learn only the 'ball gone' construction and
would not move beyond it to learn 'X gone'. What the adult needs to do is
produce other instances of the 'X gone' construction and provide enough
cues to facilitate their being understood by the child. Moreover, these
further instances need to be produced in a timely fashion: if the child is
to bring her pattern-finding skills to bear, she will still need to remember
the utterance 'ball gone' and her understanding of it when she hears 'cat
gone'.

Now consider a situation much further down the line where an adult
produces an instance of an abstract utterance-level construction that is
not at that point part of the child's inventory of constructions. In order for
this to aid the child in learning the construction, several conditions need
to be satisfied. First, the child must be able to work out the communicative intentions of the adult in uttering this sentence. This in turn will be
more likely to happen if (a) the utterance is made up largely or entirely
of words that the child knows; (b) the utterance is not overly complex
relative to the child's stage of development; and (c) the adult is prepared
to work hard to ensure that the child understands the utterance by, for
example, providing cues and repeating the utterance if it is not initially
understood.

A second condition is that the adult must resist the temptation to replace
her utterance with a simpler sentence if the child indicates that she hasn't
initially understood the target utterance. For example, suppose a child has

yet to learn the passive construction and the adult says 'the postman was bitten by a dog'. The child initially looks puzzled. Often in such cases the adult's aim is not primarily to aid linguistic development but to get across a particular message. Hence, there is a strong possibility that the adult will produce a second simpler sentence to communicate that message – for example, 'the dog bit the postman'. If this happens, the danger is that the child will turn her attention away from the initial passive to the supporting active. The upshot of this will be that she comes to associate the adult's communicative intentions with the more familiar active sentence, and so she fails to record the event in a way that will help her learn the passive construction.

A third condition for learning the abstract-level construction is that the child remembers the particular sentence produced. If all she remembers is her conclusion about the adult's communicative intentions, then she will not have recorded the data she needs to input into the pattern-finding process once further instances of the construction are encountered.

A fourth condition is that the child hears a sufficient number of instances of the abstract utterance-level construction over a relevant period of time. For example, it won't help in the learning of the passive construction if the temporal gap between hearing the first and second passive is so great that when the child hears the latter she can no longer remember the former. In such a case, the child won't have the input she needs for the pattern-finding process.

What the identification of these conditions implies is that Tomasello's learning process is a highly vulnerable one. Perhaps these conditions are routinely met but, without substantial evidence that they are, I remain sceptical. In the light of all this, perhaps the safest conclusion to reach at this point in time is that, although we have some very exciting competing theories as to how children acquire language, no decisive blow has been delivered in favour of any one of them.

6 Conclusion

In this chapter I have examined some of the most important approaches in cognitive science to the questions of the nature of our knowledge of language and how we acquire that knowledge. The most prominent of these is Chomsky's generative approach, which places a body of highly abstract syntactic rules or principles at the centre of an individual's language. Such rules have a substantial innate basis in the form of Universal Grammar. After several decades of dominance Chomsky's perspective has come under heavy attack. One line of attack involves questioning the idea that there are linguistic universals. Another is constituted by Tomasello's usage-based approach, according to which an individual's language is an inventory of constructions that are learned by employing mind-reading and pattern-finding skills that are not specific to language. My overall conclusion is that Chomsky is not fatally wounded by these attacks and that we should keep an open mind as to how we should answer the central cognitive scientific questions about language.

6 The Brain and Cognition

1 Introduction

What is the relationship between the brain and the mind or between neural states and processes and cognitive states and processes? And what are the implications of the nature of that relationship for the role of the study of the brain in cognitive science? The purpose of this chapter is to address these questions in the light of recent developments in neuroscience and the rising prominence of that discipline both in the scientific community and in the popular imagination.

2 Neuroscience and the history of cognitive science

As we saw in chapter 1, cognitive science was founded on the basis of the rejection of substance dualism and of behaviourism. Accordingly, the mind was viewed as a physical system internal to the body that mediated the connections between the world external to the body and bodily behaviour or action. Thus, for example, when a wild animal approaches a person and they respond by throwing a stone in the animal's direction, that response is mediated by processes internal to the body that serve to identify the wild animal and work out the best way to respond to it given the person's current goals. As a result of this processing, instructions are issued to the motor control systems so that the person behaves in line with their plans. It was also a foundational assumption of cognitive science that such internal processes take place in the brain, that the brain is the seat of cognition. Given this, one might have expected that neuroscience, the scientific study of the brain, would have played a prominent role in cognitive science throughout its history. However, this has not been the case for several reasons.

The first reason why neuroscience has had only a limited role in the history of cognitive science is a practical one. Neuroscientists often point out that the human brain is the most complex system in known existence, being made up of 100 billion neurons, most of which are connected to thousands of other neurons, giving rise to a system with 100 trillion internal connections. This complexity makes the human brain a very difficult system to study. Not surprisingly, our scientific understanding of the workings of the brain has historically lagged way behind that of our understanding of other aspects of the natural world. The upshot of this is that, for most of the history of cognitive science, neuroscientists haven't known enough about the workings of the brain to make much of a contribution. However, things are rapidly changing on this front, in no small part on account of the emergence of technologies that enable us to scan the human brain.

A second reason is more principled, being bound up with the computational theory of mind. Most early cognitive scientists viewed the mind as a computer that is realized or embodied by the brain. Thus, any individual human mind or cognitive process that it executes can be described in neural terms. Now suppose that we were in a position to uncover that description by successfully studying cognition at the neural level. That description would not supplant or make irrelevant the description of cognition in terms of content-bearing representations and the computational processing of them. All it would do would be to describe the lower-level details of how cognitive processes were implemented in the brain. Though valuable and interesting, such a description could not on its own explain cognition, as it would ignore the very features of our internal processes that made them cognitive and enabled them to support cognition. In effect, going entirely neural would involve changing the subject by deciding to cease to view us as cognitive systems.

This perspective isn't specific to classical computationalism. Even connectionists who complain about the lack of neural plausibility of classical computationalism are committed to the idea that connectionist networks process information and utilize representations. Accordingly, for them a connectionist explanation of cognition must highlight those representations and their contents and thus operate at a level of description higher than that of the neural.

A third reason for the limited role of neuroscience in traditional cognitive science is again principled rather than practical. One can imagine a system that can be described at various levels of abstraction where there was a very tight relationship between the levels in the respect that two distinct systems that satisfied one and the same higher-level description would also satisfy the same lower-level description. Such a relationship would hold if the categories that figured in higher-level descriptions reduced to those that figured in lower-level descriptions. In chapter 1 we encountered the type-identity theory, a theory of the relationship between the mind and the brain which pre-dated the cognitive revolution and was undermined by it. According to the type-identity theory, types of mental state are identical to types of brain state in a manner analogous to that in which water is identical to H_2O or heat is identical to mean kinetic energy (Place, 1956; Feigl, 1958; Smart, 1959). This implies that any specific mental description corresponds to a specific neural one, so that anything that the former applies to the latter will also apply to, and vice versa. However, for the type-identity theorist, such descriptions are not on the same footing, as the neural is more fundamental than the mental, and the specification of an identity relationship between a mental and a physical type (such as that pain is C-fibre firing) specifies the true nature or essence of the mental type but not the true nature or essence of the physical type.[1]

In viewing the mind as a computer, early cognitive scientists explicitly rejected the type-identity theory and endorsed a functionalist alternative that portrayed cognitive states and processes as being multiply realizable at the physical level (Putnam, 1967; Fodor, 1968b). Thus, two systems could be executing one and the same cognitive process yet be very different at the physical level. The idea of multiple realizability was taken very seriously indeed with the widespread endorsement of the idea that it is possible to program a computer with a silicon-based chemistry to engage in the very cognitive processing that we carbon-based humans routinely execute, an idea that underlay much work in Artificial Intelligence. The upshot of this view that cognitive states and processes are multiply realizable is a rejection of the idea that the essence of cognition lies at the neural level. Moreover, it was widely held, an approach to cognition that focused on the neural level would miss important generalizations, generalizations that subsumed physically different systems that ran one and the same computer program. As the aim of science is to capture generalizations, this was seen as a major failing (Fodor, 1974).

At this point in time, well into the second decade of the twenty-first century, this attitude of doubting the significance of neuroscience is widely seen as problematic within cognitive science. It is not merely that there have been technological developments that enable us to scan the brain and give us greater insights into its workings, as that alone would not undermine the second and third reasons for questioning the importance of neuroscience as described above. For there are grounds for thinking that the case made in stating those reasons is somewhat overplayed. To appreciate this, consider the following points. Recall Marr's (1982) famous and highly influential account of cognitive science, according to which a complete account of a given cognitive capacity would distinguish between three levels – the computational level, the level of representation and algorithm, and the level of implementation. On this picture, neuroscience has a key role to play in a complete cognitive science. Yet that role will be restricted to indicating how processes described at the computational and algorithmic levels are implemented in the brain. If it wasn't quite true to Marr's intentions, it was certainly the case that he was routinely interpreted as saying that the neural level was the least important level and that detailed work at that level couldn't be executed until work at the higher levels was completed. But there is an alternative way of thinking of Marr's hierarchy according to which work at the level of implementation can and should play a key role in constraining theorizing at the higher levels. Recognizing this goes hand in hand with appreciating some key elements of the computational perspective, as I will now explain.

In virtue of their commitment to physicalism, champions of the computational approach within traditional cognitive science were keen to stress that the representations they postulated are physically embodied. Moreover, the representational complexity of a representational system employed by a given cognitive system will be reflected in its physical complexity. In other words, there will be a mapping from representations onto physical states such that the complexity of any given representation will be mirrored in the complexity of the physical state it is mapped onto (Pylyshyn, 1984). Accordingly, similarities between complex representations (as when two complex representations share a particular component representation in the manner in which the English sentences 'dogs bark' and 'dogs chase postmen' share a component word) will be reflected at the physical level. But if such properties of representations are reflected at the physical level, then we should expect there to be cases where disputes between cognitive scientists operating at the higher levels of Marr's hierarchy can be resolved by appeal to data relating to lower levels. In actual fact, we do see this happening in contemporary cognitive science, and I will discuss some examples later in this chapter (appreciating the power of these examples will require having some understanding of the workings of the brain and work in contemporary neuroscience, which I will come to in the next section).

A related point is that, in talking of multiple realizability, functionalists had a tendency to exaggerate the extent to which members of a given high-level category could diverge at the physical level. Consider coffee cups, for example. Now of course coffee cups come in a range of different shapes, sizes and materials, but there are limits to this variability. For example, it is difficult to see how a coffee cup could be made out of water or sugar or something that has the shape of a blade of grass, suggesting that coffee cups might share certain abstract physical properties relating to shape, solidity, rigidity, impermeability, and the like. With respect to minds, the implication would be that a mind can't be made out of any material and that the mental complexity of a minded subject is going to be reflected in, and constrained by, its physical structure and complexity. For example, one key reason why slugs aren't as cognitively complex as humans is because they lack the required internal physical structure and complexity. But, if this is the case, then the possibility arises that information as to a system's internal physical structure and complexity can provide evidence as to its cognitive nature.

What I'm suggesting is that it is perfectly consistent with the traditional computational perspective that evidence from the neural level could be valuable in determining how our minds work at the cognitive level. However, I do want to resist a more extreme position that is driven by questioning the idea that the mental is multiply realizable.

In recent years some philosophers[2] have begun to question whether mental states are in fact multiply realizable. One argument for a reductionist perspective has been developed by John Bickle (2003) and appeals to standard practice in neuroscience. In the case of research into human medical conditions, it is common practice to experiment on animals. For example, in order to determine whether a particular food additive is carcinogenic, one might feed large doses to rats to observe what effect it has on them. Clearly, such experimentation is based on the assumption that rats are sufficiently similar to humans at the physical level so that if something has a particular effect on a rat that is grounds for concluding that it may well have the same effect on us. As in the medical case, there is a rich tradition of experimenting on the brains of non-humans in order to uncover facts about the working of the human brain. A classic example is by Hubel and Wiesel (1959), who made claims about the function of neurons involved in visual processing on the basis of experimentation on cats. But, for such work to be taken as relevant, then it must be assumed that the animals studied are relevantly similar to us. In this case such a similarity would reside at both the neural and the cognitive level. More specifically, in Hubel and Wiesel's case, the assumption is that cats, like humans, have visual experiences that represent the shapes of external objects and that those experiences are underpinned by common neural mechanisms (for example, by the firing of cells that are sensitive to lines with a particular specific orientation).

The point here is that neuroscientists proceed on the assumption that similarities between different species at the cognitive level are underpinned by commonalities at the neural level, an assumption that implies that we won't find any actual cases of cognitive states that are multiply realized. An upshot of this, argues Bickle, is that the philosophical champion of multiple realizability is adopting a view of the mind that is in tension with neuroscientific practice, something that no self-respecting naturalist should do.

I'm not entirely convinced by this argument. The neuroscientific research tradition to which Bickle refers does not involve rejecting the cognitive perspective; that is, it does not seek to describe the operations of brains purely in neuroscientific terms. Rather, it seeks to establish a bridge between the neural and the cognitive by revealing how cognitive processes described in cognitive terms are underpinned by neural processes. The work also involves a commitment to the role of evolution in the neural domain; our brains, just as much as any other biological system, are the product of evolution by natural selection. It is because of this that we should expect there to be similarities between the human brain and the brains of our simpler relatives. And such similarities will give rise to cognitive similarities. Given all this, we should expect that all naturally evolved minds that are similar at the cognitive level will be similar at the neural level, and hence that there will be few actual examples of multiple realizability (that is, cases where two creatures share a cognitive state while being very different at lower levels). The key point, then, is this: the apparent absence of cases of multiple realizability is not a product of the fundamental nature of the cognitive or the mental but a product of contingencies relating to the emergence of cognition and mentality in the actual world. In sum, then, there is nothing about neuroscientific practice that serves to undermine the traditional idea that cognitive states are multiply realizable at the physical level.

What I have argued in this section is that we should expect neuroscience to have a positive role to play in cognitive science, one that goes beyond the role that it played in the early years of the discipline. To say this does not involve rejecting the computational perspective either in its classical or in its connectionist version. But it is not merely that knowledge about the workings of the brain at the neural level can help us to understand human cognition as the relationship of evidence can run in the opposite direction. Suppose that someone wanted to understand the workings of the brain at the neural level and had no direct concern with whatever cognitive processes or capacities such workings underlay. I doubt that in practice they would make much progress. Put crudely, the activity of the brain is so vast and complex that they wouldn't be able to make sense of much of that activity. In reality, neuroscientists have to view humans as cognitive agents in order to gain a route of access into the brain. This is clear in the case of brain-scanning experiments where subjects are asked to carry out a particular cognitive task while having their brain scanned. In such experiments the neuroscientist is assuming that the subject understands the instruction and succeeds in carrying out the relevant cognitive task, be it thinking about a particular event, carrying out a mathematical calculation, or identifying an object. Without viewing the subject as a cognitive agent who succeeds in carrying out a particular cognitive task that they understand themselves as having been requested to execute, it is difficult to see what we could learn from brain-scanning experiments. Even in those cases where non-human animals are involved, it is clear that a cognitive perspective is taken. For example, Rizzolatti and his colleagues (1996) famously postulate the existence of mirror neurons in the pre-motor cortex of monkeys, neurons that fire both when a particular goal-directed action is carried out and when an instance of the same action is observed. Identifying such neurons as mirror neurons presupposes interpreting the monkeys on one occasion as acting in the service of a particular goal and on another occasion as understanding the behaviour of one of their fellows as being an instance of that very goal-directed behaviour.

The discussion so far has been relatively abstract. What I will do in the remaining sections of this chapter is turn to a more concrete discussion of neuroscientific findings and their significance within cognitive science. In the next section I will provide a general account of the structure and workings of the brain. In sections 4 and 5 I will focus on the neuroscience of vision. The reasons for this are threefold. First, the study of vision is perhaps the most advanced aspect of neuroscience. Second, focusing on the study of vision will enable the discussion to dovetail with the earlier discussion of the foundations of cognitive science in chapters 1 and 2, where Marr's theory of vision was very prominent. Third, visual perception has been of perennial concern to philosophers, those working both in the philosophy of mind and in epistemology. Hence, by focusing on the neuroscience of vision we can connect the discussion to some traditional philosophical concerns.

3 The basics of the brain

The brain is part of the nervous system, and that system has two components – namely, the central nervous system and the peripheral nervous system. The central nervous system consists of the brain and the spinal cord and the peripheral nervous system consists of the nerves that serve to connect the central nervous system to muscles and sense organs. The peripheral nervous system can in turn be broken down into the somatic and automatic nervous systems. The somatic nervous system is involved in the voluntary control of the skeleton. The automatic nervous system is involved in homeostasis – that is, the regulation of internal bodily conditions beneath the level of conscious control. Thus, the automatic nervous system is involved in the control of breathing, digestion, internal temperature, heart rate, sexual arousal, and so on.

Like all complex biological systems, the brain is made up of cells,[3] the cell being the most basic biological unit that – at least in principle – can survive in isolation. Thus, cells are the basic units of life. There are different types of cell in the brain, but the type that receives most attention and is most directly involved in cognition are nerve cells or neurons.[4] There are approximately 100,000 billion neurons in a normal human brain. Each of these neurons connects with many other neurons, giving rise to a network of approximately 100 trillion connections. In virtue of the colossal size of these numbers, it is routine for the brain to be described as the most complex structure in the known universe.

There are many different types of neuron, and one broad way of classifying them relates to their function. Thus, motor neurons carry signals from the central nervous system to muscles, so driving bodily movement.

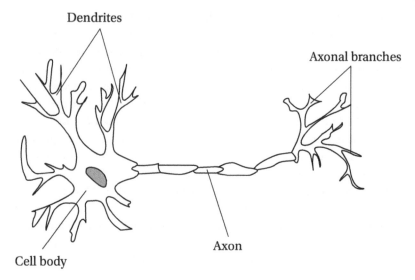

Figure 6.1   A neuron

Sensory neurons carry signals from the sensory systems to the central nervous system. And interneurons connect neurons in the central nervous system with other such neurons, facilitating communication between them. This variety of function explains why neurons can differ so much in their shape and size; for example, a neuron that carries signals from the central nervous system to the foot, so controlling its movement, might be over a metre in length, whereas some interneurons are less than 1 millimetre across.

Despite their differences, all neurons have much in common, reflecting their communicative behaviour. They have a cell body containing a nucleus and other organelles that keep the cell alive. Leading to the cell body are numerous root-like structures, known as dendrites, that carry stimulation in the form of a passive electrical current to the cell body (see figure 6.1). This stimulation is the neuron's input and is the product of external factors impinging on the neuron. Each dendrite carries stimulation that originates from a different source, such as another neuron, a muscle cell, or energy impinging on a sensory surface such as the retina. Leading away from the cell body is the axon. The axon carries the signal that is the neuron's response to the input delivered by its dendrites. The axon itself splits into many branches, so that the signal it carries can be passed on to many distinct structures, such as other neurons or muscle cells. Thus, neurons transform input signals into output signals. In doing this, the neuron must 'decide' whether the input delivered by its many dendrites merits delivering a signal as output. This typically involves a process of summation where the neuron calculates whether the sum of its input signals meets a certain pre-determined threshold. If this threshold is met, then the neuron will 'fire' a signal down its axon; if not, then it remains silent.

What I have been describing as 'signals' take an electrical form. The signals carried by the dendrites are passive currents, whereas those carried by axons are active electrical impulses known as action potentials. Action potentials are constituted by waves of movement of positively charged ions of sodium and potassium into and out of the neuron. Thus, when a neuron fires in response to external stimulation, it generates an action potential that travels along its axon. This action potential is an all or nothing matter and so is binary in nature.[5] Thus, the individual action potentials generated by a neuron don't vary in strength or significance. Nevertheless, a neuron can carry signals of varying strength or significance, as it can generate a range of different patterns of action potentials (for example, the higher the rate of firing, the greater the strength of the signal).

Action potentials are generally conceived of by neuroscientists as carrying information. What information a particular impulse or pattern of impulses carries is not determined by its intrinsic nature; thus, a firing in one part of the brain might carry quite different information than a physically identical one elsewhere in the brain. Rather, the identity of the information has to do with such relational factors as what typically causes the neuron in question to fire, what effects its firing typically has, what biological function the firing has, and such like. For example, consider a neuron that typically fires when the visual system is stimulated by a short line with a 45 per cent orientation. Then that firing might carry the information that the distal scene features a line of that length and orientation. Of course, much more can be said on this issue; indeed, there is a rich philosophical literature concerned with explaining how mental representations come to mean what they mean or have the content that they have. Such theories can be explicitly pitched at the level of action potentials, as it is natural to hold that action potentials constitute the neural realization of at least some mental representations.[6]

Although the signals carried within neurons take an electrical form, neurons generally stimulate – and so communicate with – other neurons by chemical means. The point at which the axonal branches (that is, the fibres emanating from an axon) of one neuron meet with the dendrites of another is called a synapse. There is a small gap between the axonal branches and the dendrites with which they connect that the action potential cannot jump over. This is known as the synaptic cleft. Hence, neurons must find some non-electrical way to communicate with one another. The solution they have evolved involves chemicals known as neurotransmitters. When the action potential reaches the terminus of an axonal branch it causes the release of a jet of neurotransmitter, which crosses the synaptic cleft and binds to receptors on the receiving dendrites. The effect the neurotransmitter has can be either excitatory or inhibitory. In the former case, the neurotransmitter pushes the receiving neuron towards firing and, in the latter case, it pulls it away from firing. There are numerous distinct neurotransmitters used by the brain and some of them have very familiar names, such as 'dopamine' and 'serotonin'.

The neurons in the brain do not form an undifferentiated mass; rather, they are organized into circuits that are in turn grouped into higher-level units, and so on right up to tissue structures that would be visible to the naked eye if a brain was removed from its encasing skull.[7] I will now turn to describing the organization of the brain at the highest levels.

The brain is composed of a number of distinct units that are different from one another in both their appearance and function. Some of these units are more recent in evolutionary terms than others and are either absent or not developed to the same extent in non-human animals. Others are much older and have direct analogues that can be found in the brains of many cognitively simpler animals.

Consider a typical imagistic representation of the brain. This will portray an entity not unlike a shelled walnut in appearance that has a distinctive rippled outer surface. This outer surface is known as the cerebral cortex and is a sheet of grey matter some 2 to 4 millimetres thick. The term 'grey matter' relates to the colouration of the cerebral cortex, which reflects the fact that it is made up largely of cell bodies. Grey matter is to be distinguished from white matter, which is brain tissue that is made up largely of axons and has a whitish appearance, because axons are often coated with a fatty substance known as myelin in order to aid the transmission of action potentials along their length. Were the cerebral cortex laid out flat, it would cover an area of approximately 1,350 cm^2. However, it is repeatedly folded to enable it to fit into the skull, so forming troughs known as sulci and peaks known as gyri.

The brain as a whole divides into two halves, the hemispheres; by and large, structures in one hemisphere are replicated in the other and perform closely related functions, though there are some notable exceptions. This hemispheric division applies to the cerebral cortex, each half of which is divided into four lobes, named after the bone under which they are located (see figure 6.2). Different brain structures, including each of the cerebral lobes, have distinct functions. Although these functions are typically characterized in terms of the distinct cognitive and behavioural capacities they support, it is important to realize that any such characterization is likely to be only a crude approximation of the truth for three reasons. First, each distinctive structure in the brain constantly interacts with other structures in doing what it does. Second, each structure can play a role in supporting what from the commonsense perspective are quite different cognitive and behavioural capacities. And, third, the exercise of most such capacities will involve the operation of more than one structure.

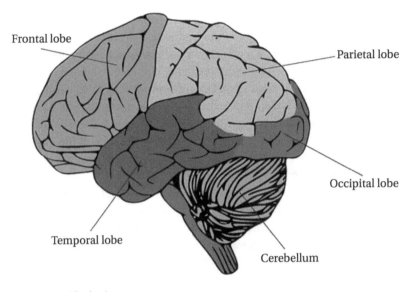

Frontal lobe

Parietal lobe

Occipital lobe

Temporal lobe

Cerebellum

Figure 6.2   The brain

The cerebral cortex is divided into four areas or lobes, each with the following rough functions. The frontal lobes run from the front of the brain to a
deep groove known as the central sulcus located half-way along the top of
the brain. Its primary function relates to motor behaviour – that is, to planning what actions to carry out and generating the instructions that initiate
those actions. Different areas of the frontal lobe make their own specific
contribution to this general motor theme. The area at the very front of the
frontal lobe (the most 'anterior' area in neuroscience speak) is concerned
with the most abstract aspects of planning. An example would be the
process of your deciding to stop reading shortly and go and make a cup of
coffee. The prefrontal lobe takes up most of the frontal lobe and is far more
developed in humans than in any other species, explaining our peerless
capacity to plan and engage in flexible behaviour. At the rear (or posterior)
part of the frontal lobe is a strip known as the primary motor cortex. It
generates the specific instructions that are sent directly to motor neurons
which, in turn, connect to the muscles controlling bodily movement. It
would thus be involved in causing the specific bodily movements that you
carried out when you made that cup of coffee. One very interesting feature
of the primary motor cortex is that it encodes what is known as a motor
map. That is, specific parts of the primary motor cortex correspond directly
to specific parts of the body in the respect that their activation causes
movement of that body part. Moreover, adjacent parts of the body are
caused to move by activity in adjacent parts of the primary motor cortex.
There are several other examples of maps encoded in the brain.

   Lying behind the frontal lobes are the parietal lobes. They are involved in bodily sensation – for example, in the perception of pain, pressure, temperature, taste and touch. Just as the frontal lobes contain a motor map, the parietal lobes contain a map of the skin. That is, different areas of the lobes are directly concerned with processing input from different parts of the skin, with adjacent areas of this map corresponding to adjacent areas of the skin. This map is distorted in the respect that a larger area is devoted to parts of the skin with high sensitivity – such as those on the face and the fingertips – than those of low sensitivity – such as the soles of the feet and the back of the neck. This distortion explains why some parts of the skin are more sensitive to stimulation than others.

The occipital lobes are located at the back of the brain. These lobes are concerned entirely with visual processing and take their input from the retinas, though they do pass some of their output to the temporal and parietal lobes for higher-level visual processing (for example, that involved in object recognition). Within the occipital lobe there are different areas, known as V1, V2, V3, V4, and so on, that carry out different aspects of visual processing – for example, relating to motion detection, depth perception, colour perception, and so on. In other words, the visual processing task is broken down into a series of subtasks each executed by a distinct sub-unit of the occipital lobe.

The temporal lobes lie along the curved side of the brain above the ears and play a role in a number of functions, such as the storage of visual memories. They take input from the occipital lobes to compare current visual input with those memories in order to facilitate object recognition. The temporal lobes are also involved in auditory processing, taking input from the ears, and the left lobe contains an area known as Wernicke's area that plays a key role in language comprehension.[8]

Beneath the cerebral lobes the brain is made up largely of white matter and four fluid-filled cavities known as ventricles.[9] However, there are a number of sub-cortical grey matter structures. The diencaphalon lies immediately below the cerebral cortex and consists of the thalamus and the hypothalamus. The former is often described as a relay station, as it directs input coming from the sense organs to the relevant part of the brain and directs the output of brain regions to other parts of the brain, so facilitating communication within the brain. The hypothalamus, on the other hand, controls homeostatic functions such as body temperature, the sleeping–waking cycle, blood flow, hunger and thirst, the level of sex hormones, and such like.

Next we come to the limbic system, which is a collection of structures, including the amygdala and hippocampus, that are very old in evolutionary terms, being present in such animals as reptiles and birds. The limbic system plays a key role in memory and emotion. It is involved in both the recognition of the emotional states of others and in generating our own emotions. It also enables connections to be made between our memories and emotions so that some of the events we remember have an emotional significance to us.

The basal ganglia comprise a group of tightly related structures, namely, the caudate, the putaman, the globus palladus, the substantia nigra and the subthalamic nucleus. Via interaction with the thalamus and the frontal lobes of the cerebral cortex, the basal ganglia play a key role in the planning, coordination and execution of bodily movements.

The brainstem marks the point of transition between the spinal cord and the brain and, in ascending order, consists of the medulla, the pons and the midbrain. These are the oldest parts of the brain in evolutionary terms, being present in most vertebrates. They are involved in the control of such basic behaviour as eye movement, in reflex actions such as swallowing and vomiting, and in homeostatic functions such as the regulation of respiration, blood pressure, heart rate and body temperature.

Finally we come to the cerebellum, which lies directly behind the brainstem and is not covered by the cerebral hemispheres. Once more, this is an old structure of the brain that contains more than half of the total number of its neurons. The function of the cerebellum relates to motor coordination, and it enables us to keep our balance and execute smooth, non-jerky movements.[10]

This account of the brain emphasizes the point that different parts of the brain have different evolved functions that are common across all human individuals. However, some neuroscientists are keen to emphasize the plasticity of the brain, the fact that the brain can change significantly due to the contingencies of experience. One element of such plasticity is learning, and a prominent view is that the brain engages in Hebbian learning, a process named after the Canadian psychologist Donald Hebb (1949). This is commonly captured by the slogan 'what wires together fires together'. The basic idea is that, when two neurons repeatedly fire at the same time, the synaptic connections between them strengthen so that the firing of one will cause the firing of the other. Alternatively, as Paul Churchland (2007, 2013) characterizes Hebbian learning, when a number of neurons simultaneously fire, the synaptic connections between those neurons and any downstream neuron to which they all connect are strengthened.

However, when some neuroscientists describe the brain as plastic, they mean more than that the brain changes through everyday learning. For example, parts of the brain that normally receive one type of input can, when starved of that input, turn to process a different type of input. V. S. Ramachandran (2011) documents such a case. Some individuals who have had an arm amputated experience sensations where their arm would have been as a result of stimulation to their face. What explains such 'phantom limb' cases is that that part of the parietal lobe normally involved in processing signals from the arm responds to a period of dormancy by processing signals resulting from stimulation of the face. Hence, such facial stimulation leads to sensations relating to the amputated arm as well as the face.

Ramachandran takes the case of phantom limbs to be of great significance. He writes:

> Generations of medical students were told that the brain's trillions of neural connections are laid down in the foetus and during early infancy and that adult brains lose their ability to form new connections. . . . Our observations flatly contradicted this dogma by showing, for the first time, that even the basic sensory maps in the adult human brain can change over distances of several centimetres. (2011: 28)

However, I think we should be wary of reading too much into such cases. For example, I don't think they constitute a clear-cut case of one region of the brain taking on a new function for the following reason. The portion of the parietal lobe that normally receives signals from the arm is located next to that which receives signals from the face, so what happens is that, so to speak, activity in one part of the parietal lobe is communicated to, or spills over into, an adjacent part. This hardly counts as a case of fundamental rewiring. Moreover, once stimulated by signals from the face, the area that historically processed signals coming from the arm continues to give rise to experiences related to the arm; it doesn't, for example, give rise to facial sensations. Thus, the representational significance of activity in this area doesn't change.

The level of understanding of the basic structure and functioning of the brain described so far in this section has been hard won, and this is hardly surprising given the complexity of the brain and the ethical obstacles standing in the way of brain dissection and surgical interference with the brains of living subjects for research purposes. But, in recent years, great strides have been made as a result of the development of brain-imaging technologies, which enable neuroscientists to observe the structure and functioning of living brains in relatively safe and non-invasive ways. Here are some of the most important such technologies.

A basic distinction is that between techniques that uncover the long-term, fixed structure of the brain – structural imaging techniques – and those that detect short-term changes associated with neural activity – functional imaging techniques. Two of the most important structural imaging techniques are Computerized Tomography (CT) scanning and Magnetic Resonance Imaging (MRI). CT scanning involves taking X-rays of the brain and is particularly effective at identifying the size and location of tumours and lesions, but it does not register the difference between grey and white matter and so is of limited value in identifying the key structural units of the brain. MRI relies not upon X-rays but, rather, on applying a strong magnetic field to the brain and then delivering brief radio-wave pulses. This serves to align and then displace protons in the water molecules in the brain, thereby producing an MR signal. MRI distinguishes between grey and white matter and is superior to CT scanning in terms of spatial resolution – that is, it provides more fine-grained spatial detail.

The two most prominent functional imaging techniques are functional Magnetic Resonance Imaging (fMRI) – a derivative of MRI – and Positron Emission Tomography (PET). Both of these are based upon the idea that neurons require oxygen provided by the blood in order to survive, so that an increase in their activity will require an increase in the blood supply to them. In the case of fMRI, increases in blood supply to a particular region of the brain are measured by detecting changes in blood oxygen levels in that region. An increase in blood supply to support increased neural activity will be associated with an initial dip in oxygen levels as oxygen is consumed, followed by a rise in oxygen levels as fresh oxygenated blood arrives. In PET a radioactive tracer is injected into the bloodstream so that the most active areas of the brain emit the strongest radioactive signal.

In order to determine which areas of the brain or collections of neurons are centrally involved when carrying out a particular cognitive task, both fMRI and PET require the comparison of images gained from scanning the brain when the task in question is being executed with those gained from scanning the brain when it is not carrying out this task. This technique is sometimes known as cognitive subtraction, and an example will be provided in the next section in connection with some important work on vision.

fMRI has some key advantages over PET, which explains why it has become the dominant form of functional imaging. As radioactivity is not involved, subjects can be scanned many times rather than just once. fMRI also offers superior spatial and temporal resolution – that is, finer-grained detail as to the spatial and temporal location of neural activity.

4 Vision

Visual perception enables us to discover a rich body of information about the world lying beyond our outer surfaces, information that can be used to guide our actions so that we satisfy our needs and desires. Visual perception has been extensively studied by cognitive scientists and is arguably the best understood element of our cognitive lives. In this section I will provide an account of current orthodoxy in vision science along with some more controversial recent developments. This is of particular relevance to this chapter, as the study of vision has been that area of cognitive science where developments in our understanding of the brain have played the most prominent role.

On the one hand, vision is a very familiar phenomenon. We open our eyes and direct them at the outer world, and without any conscious effort we have a rich experience in which we see a three-dimensional world populated by objects of various shapes, sizes, colours and surface textures standing in spatial relations to one another. Our visual experiences rarely stand still; rather, they change from one moment to the next. However, this dynamism doesn't generally involve sudden leaps and breaks in our experience; our experience exhibits a flow and continuity so that vision has a temporal dimension in that we see particular objects as moving over time and following unbroken paths when they move (Bayne, 2010).

The preceding paragraph describes how vision seems to be to us from an ordinary everyday perspective. But even if a cognitive scientist didn't wish to quibble with that take on vision, she would certainly argue that there is a lot more to be said about the visual process. What exactly is going on in the mind-brain that enables us to see the external world in the way in which we see it? Commonsense has little to say in answer to this question, but when cognitive scientists have addressed it they have come up with answers which are often very surprising from a commonsense perspective, as we will soon see.

The visual process begins with light reflected off external surfaces and objects being focused onto the retina, a collection of roughly 125 million light-sensitive cells at the back of each eye. The retina is a transducer which converts light energy into an electrical signal that can be processed by the mind-brain. Retinal cells fall into two broad categories, namely rods and cones. Rods are sensitive to low-intensity light and are distributed evenly across the retina. Cones are sensitive to the different wavelengths of light that correspond to the colours we perceive. Cones are most densely packed in an area of the retina known as the fovea, and it is this area that is central both to colour vision and to the perception of fine detail in the viewed scene. When stimulated by a viewed scene, each cell of the retina will correspond to a particular location in that scene, and adjacent cells will correspond to adjacent external locations. Such a correspondence is pre-served as the signal is passed on through further layers of neurons, giving rise – once again – to the kind of maps that we saw earlier in the general description of the brain's workings.

The output of the rods and cones is fed to a layer of neurons known as the retinal ganglion cells, and their axons form the optic nerve which carries signals from the eye to the brain. Each retinal ganglion cell has a receptive field which is a matter of the number of adjacent retinal cells that feed it its input. Ganglion cells that take input from rods tend to have larger receptive fields than those that take their input from cones.

Once it has left the eye, the information carried to the brain can follow three distinct routes or pathways. In humans the most important pathway is the geniculostriate pathway, which passes through an area of the thalamus known as the lateral geniculate nucleus (LGN for short). The LGN in the left hemisphere receives input from the right-hand side of the viewed scene and the LGN in the right hemisphere receives input from the left-hand side of the viewed scene. As both sides of the viewed scene stimulate the retina of each eye, each LGN receives input from both eyes. This is possible because the optic nerves coming from each eye cross just before entering the brain at a point known as the optic chiasma. Here there is an exchange of information, with signals emanating from the left-hand side of each retina (which are stimulated by the right side of the viewed scene) being directed to the left hemisphere and the signals emanating from the right-hand side of each retina (which are stimulated by the left side of the viewed scene) being directed to the right hemisphere.

The receptive fields of the LGN cells have centres and surrounds, the centre corresponding to a particular point on the retina (which in turn corresponds to a particular point in the viewed scene) and the surround corresponding to the immediately surrounding points on the retina. LGN cells respond to points of light. Some fire when the centre of their receptive field is stimulated while the surround is not, whereas others respond when their surround is stimulated while the centre is not (stimulation to both the centre and surround cancel each other out, so that an LGN cell does not fire strongly in such cases). Given this, it is natural to think of the firings of LGN cells as representing the presence and location of spots of light falling on the retina.

The output of the LGN cells is then carried to the primary visual cortex, an area of the occipital lobe that is also known as V1. In their pioneering work in the 1960s David Hubel and Torsten Wiesel (1962) discovered that neurons in V1, rather than responding to points of light falling on the retina, respond to bars of light of specific length and orientation. They do this because each V1 neuron ultimately receives its input from a particular column of retinal ganglion cells and so will fire strongly only when the cells making up that column fire strongly while the surrounding cells outside of the column do not. The detection of such bars of light is important for further downstream visual processing, as they typically correspond to salient features of objects in the viewed scene such as edges and boundaries. It is natural to view the strong firing of V1 cells as representing the presence and location of bars of light of specific length and orientation on the retina.

On leaving V1, visual information follows two distinct pathways. On the one hand, the so-called ventral stream runs from V1 through V2 and V3 and on to the temporal lobe. On the other hand, the dorsal stream runs through V5 and V3 and on to the parietal lobe. Ungerleider and Mishkin (1982) famously argued that these distinct pathways have quite distinct functions. They hypothesized that the ventral stream is concerned with object identification, and so dubbed it the 'what pathway', and that the dorsal stream is concerned with object location, and so dubbed it the 'where pathway'. Subsequently, Goodale and Milner (1992, 2004) developed the highly influential view that the function of the dorsal stream is to enable us to act on objects, and so they called it the 'how stream' or the 'vision for action pathway'.

The postulation of such functionally distinct visual processes might appear counter-intuitive, as it doesn't appear to be part of our conscious experience. I am certainly not consciously aware of executing distinct processes when I identify an object as a coffee cup and work out what movements I have to make to grasp it and lift it to my lips. This brings out two important points. First, much of what goes on in visual processing is unconscious. Indeed, there is a well-known phenomenon called blindsight (Weiskrantz, 1986; Cowey, 2004) where individuals with little or no conscious experience display behavioural evidence which suggests that they can detect certain visual stimuli. Second, there is much functional specialization in the brain, with distinct areas of the brain carrying out distinct tasks. Often, such specialization is surprising from a commonsense perspective. For example, there is considerable evidence that identifying faces is a process distinct from identifying objects in general and is carried out by a dedicated area of the brain known as the fusiform face area (Bruce and Young, 1986).

If the functionally distinct processes involved in visual perception reside at a level beneath conscious awareness, then the question arises as to what evidence we have for their existence. One traditional source of evidence comes from cognitive neuropsychology. This is the study of individuals who suffer from some specific cognitive deficit due to localized damage to their brain brought about by a genetic defect, a stroke, a blow to the head, or whatever. Suppose we have two cognitive tasks, C1 and C2, and the question is whether or not these two tasks are subserved by one and the same set of cognitive processes and representations. Suppose that we found individuals who could carry out C1 while being unable to carry out C2 and other individuals who could carry out C2 without being able to carry out C1. Then the tasks would be doubly dissociable. Such a double dissociation would constitute evidence that C1 and C2 did indeed rely upon distinct processes and representations.

Goodale and Milner (2004) appeal to double dissociations in order to justify their postulation of two distinct visual pathways. On the one hand, they argue, there is the vision for perception pathway, which issues in conscious visual experiences and is implicated in object recognition on the basis of shape and form. On the other hand, there is the vision for action pathway, which is concerned with guiding action. They document the case of Dee Fletcher, who, as a result of carbon monoxide poisoning, suffers from visual form agnosia. Although she is not totally blind in the sense of having no conscious visual experiences, her vision is severely limited. She can see colours and textures, but she cannot perceive the shape and form of what she sees or the orientation of lines. So, for example, she cannot see the difference between a triangle and a square or a horizontal and a vertical line. As a result she has problems recognizing objects. Nevertheless, she is able to interact with objects quite successfully. For example, she can grasp and manipulate a pencil she is offered, and in experiments she can orientate a card and push it through a slot even though she can't consciously see the orientation of the slot. In short, despite her perceptual problems, she is able to use visual information to guide her actions.

Individuals who suffer the opposite dissociation, optic ataxia, have damage to the parietal lobe in both hemispheres. They don't have the perceptual problems of Dee Fletcher as they are able consciously to see the size, location and orientation of objects, but they have problems with respect to action guidance – problems with successfully reaching and grasping for objects they can consciously see.

Though historically important, there is a problem with relying heavily on evidence of double dissociations between cognitive tasks. Suppose that no individuals are found who are capable of carrying out C1 and not C2 and, similarly, no one is found who is capable of carrying out C2 and not C1. It doesn't follow from this that the two tasks depend upon the same processes and representations so that they are not distinct in any fundamental sense. For it might be that the distinct processes and representations are located close to one another in the brain, so that the neural damage (say, a lesion caused by a stroke) that 'knocks out' one set tends to 'knock out' the other. Factors such as this, alongside the relative scarcity of individuals with relevant neural damage for cognitive scientists to study,[11] place limits on the fruitfulness of a cognitive neuropsychological approach.[12]

Another important source of evidence is provided by brain scanning. Suppose that in an experiment a subject is instructed to carry out task C1 while having their brain scanned. It is found that, while they execute C1, a particular region of their brain (call it B1) is particularly active – a region that was inactive both before the execution of the task and after its completion – while another region (B2) remains inactive. In addition, when they execute C2, B1 is inactive while B2 becomes active. On the assumption that distinct processes and representations are located in different parts of the brain, such evidence would support the claim that C1 and C2 do indeed depend upon distinct processes and representations.

Evidence from brain scanning has been presented by Goodale and Milner (2004) in order to support their postulation of two distinct visual pathways. For example, fMRI scanning has indicated that an area in the temporal lobe known as the lateral occipital area (LO) is particularly active when we look at everyday objects. Such scanning involves showing regular pictures of objects and scrambled pictures of those objects. The latter are subtracted from the former so as to take away the brain activity relating to the perception of the constituent lines and edges, thereby leaving that activity relating to the perception of the structure residing in the pictures of the intact objects. Scanning of Dee Fletcher reveals that her lesion closely matches the LO area in normal subjects with respect to its extent and location. This leads Goodale and Milner to state that

> The brain imaging experiments and clinical studies both point to one undeniable conclusion: our perceptual experience is not the product of a general purpose object recognition system but is instead the creation of a set of quasi-independent visual modules. (2004: 62)

The role that evidence from brain scanning can play when deciding between competing cognitive theories rather undermines the perspective associated with Marr (1982). According to this perspective, one can and should study cognition at a level of abstraction that ignores details of the brain's workings, and such details are relevant only to the question of how cognitive processes are implemented at the hardware level. The alternative perspective, which views the study of cognition as being inextricably linked with the study of the brain, has become dominant in recent years and is central to cognitive neuroscience.

The characterization of the visual process that I have given so far might appear to portray vision as a bottom-up process – that is to say, a process where the visual system processes retinal input uninfluenced by factors such as the individual's contingent knowledge or beliefs or their specific goals and interests at the time of retinal stimulation. If vision is bottom-up in this way, then the output of the visual process will be determined by its input in the respect that, for the output to have been different from what it actually was, the input would have had to have been different.[13] However, in recent years the idea that there are top-down factors involved in vision has become increasingly influential. To see this, consider the following cases.

Talk of cones as being retinal cells that are implicated in colour vision and as being sensitive to the light of specific wavelengths might give rise to the impression that the colour we perceive a surface or object as having is exhaustively determined by the wavelength of the light that is reflected from it onto the retina. Now, although wavelength of light does matter with respect to perceived colour, it is not the only factor involved. Consider the familiar phenomenon of colour constancy. As objects move around the world, lighting conditions change. Accordingly, the wavelength of the light reflected off a particular surface can change significantly over time. Now suppose you arrange to meet a friend on a bright sunny day at some outside location and you see that they are wearing a shirt that is a particular shade of red. Shortly afterwards you both go to a dimly lit café. In the café the wavelength of light reflected off your friend's red shirt onto your retina will be somewhat different from what it was when you were outside in the bright sunlight. Nevertheless, the colour of the shirt will not have appeared to you to have changed; rather, your colour experience will have been constant through the change in lighting conditions. The standard explanation for this is that your visual system 'knows' that surfaces do not routinely change their colour and that a particular surface can reflect light of quite different wavelengths without having changed its colour as a result of changes in lighting conditions. Thus, the visual system compensates for the changes in wavelength so that the appearance of the shirt with respect to colour remains constant (Zeki, 1993).

Perhaps a more dramatic and surprising example relates to the role shape can play in influencing the perception of colour. In an interesting experiment subjects were shown photographs of fruit that had been adjusted to be grey in colour (Hansen et al., 2006). Rather surprisingly, when the subjects were shown a picture of a grey banana they indicated that the banana looked yellow to them. What seems to be going on here is that the visual system identifies the pictured object as being a banana on the basis of its shape and, in compensating for the 'grey' light stimulating the retina, draws upon its 'knowledge' that bananas are normally yellow. In other words, it endorses the hypothesis that a yellow object is being viewed in murky lighting conditions in preference to the hypothesis that a grey one is being viewed in normal lighting conditions. And the endorsement of this hypothesis affects the conscious experience of the colour of the picture.

Other prominent cases of top-down factors involved in perception relate to attention. Attention is the phenomenon of focusing on something in particular to the exclusion of other potential objects of focus. It is nicely characterized in the following well-known and oft-quoted passage by William James, the American pragmatist philosopher and founding father of psychology in the English-speaking world:

> Attention is . . . the taking into possession of the mind, in clear and vivid form, of one out of what seem several simultaneously possible objects or trains of thought. Focalization, concentration, of consciousness are of its essence. ([1890] 1995: 403–4)

There are two closely related phenomena connected to attention. The first is called inattentional blindness. Simons and Chabris (1999) conducted the following experiment.[14] They showed student subjects a video of a basketball game played between a team dressed in black and a team dressed in white and asked them to count the number of passes made by the players wearing white. In the course of the video a woman dressed in a gorilla suit walked across the court and, while directly facing the camera, beat her chest with her hands. At the end of the video the subjects were asked if they had seen anything unusual when they were watching the basketball game. Despite her being on screen in a prominent location for a full nine seconds of the one-minute video, half of the subjects claimed not to have seen the gorilla or anything else unusual. What explains this surprising result? The input to the visual system is so potentially rich in significance that to be fully processed would be to place a massive burden on the mind-brain. In order to avoid processing overload, decisions have to be made as to which aspects of the input to process and which, in effect, to ignore. In the case of the basketball game, the subjects have the goal of counting the number of passes made by players wearing white and so attend to those players. Consequently, their visual systems focus their processing energy on those aspects of the visual input most directly related to this goal, ignoring other elements of the scene that are deemed likely to be irrelevant. The moral of this experiment is that what we attend to plays a distinct role in how we perceive a viewed scene and that the direction of our attention can be shaped by our contingent goals and interests. Thus, a particular pattern of retinal stimulation could give rise to quite different perceptual experiences as a result of attentional differences due to divergent interests and goals.

The second phenomenon is called change blindness, where quite dramatic changes in a viewed scene go unnoticed because attention is not being directed at those aspects of the scene. For example, in one experiment[15] subjects watched a video of a couple having a conversation in a restaurant. Part way through the video the plates on the couple's table changed from red ones to white without this change being noticed by the subjects.

What the phenomena of inattentional and change blindness suggest is that how an individual views a given scene is not exhaustively determined by the nature of the scene or how it affects their retina. Rather, it is influenced by which specific aspects of the scene the individual attends to, something that is influenced in turn by their contingent goals and interests.

Another top-down factor involved in visual processing is suggested by an increasingly influential body of cognitive scientists who argue that the mind-brain applies Bayes's theorem in the course of cognitive processing.[16] To see what this comes to, we have to understand Bayes's theorem, a theorem developed by the eighteenth-century English clergyman Thomas Bayes.

Bayes's theorem has to do with probability and how one should revise one's beliefs in the light of new information. The relevant notion of probability is subjective probability, which has to do with how confident or certain one is that a particular proposition is true.[17] Suppose that it is 9.00

a.m. on a June morning and the proposition in question is that it will be raining at 12.00 noon in Oxford (call this proposition p). Suppose that I am in London some 50 miles away and have received no recent information concerning the weather conditions in Oxford. What will my view of proposition p be? If asked, I will conclude that it is unlikely to be true, but I will not be absolutely certain that it is false. My degree of confidence in the truth of the proposition can be represented mathematically by a number between 0 and 1: if I'm absolutely certain that the proposition is true then its subjective probability for me at that time will be 1, and if I am certain that it is false the subjective probability will be 0. If I think that p is just as likely to be true as false then the subjective probability is 0.5, and so on. With respect to p, I reason that in the summer it rains in Oxford at midday about once every ten days and so settle on a subjective probability of 0.1.

Just as we can talk of the probability of a proposition, we can talk of the conditional probability of that proposition given some other proposition. For example, suppose that q is the proposition that it is cloudy and overcast in Oxford at 9.00 a.m. The conditional probability of p given q is a measure of my degree of certainty that p is true on the assumption that q is true. Suppose that this is greater than 0.1 because I believe that most summer days where it is cloudy and overcast at 9.00 a.m. are days where it is raining at 12.00 noon. More specifically, suppose that this conditional probability is 0.6. I then ring a friend in Oxford at 9.00 a.m., who tells me that it is cloudy and overcast in Oxford and, taking him to be honest and reliable, I become certain that q is true. Then the rational thing to do would be to revise my degree of certainty in the truth of p in line with the conditional probability of p given q. As that conditional probability is 0.6, the rational thing to do would be for me to come to believe that p with a certainty of 0.6. To continue to believe that p with a certainty of 0.1 would be irrational, as I have received new information that, given my prior commitments, I regard as relevant to the probability of p's being true.

But if rationality requires us to revise our beliefs in the light of new information in this way, then the question arises of how we are to work out the conditional probability of a proposition given another proposition. This is where Bayes's theorem becomes relevant, as it is an equation for working out the conditional probability of a proposition p given another proposition q. It can be stated in the following terms:

The conditional probability of p given q equals the probability of q given p multiplied by the probability of p divided by the probability of q.

In more abstract terms:

$$Pr(p/q) = \frac{Pr(q/p) \times Pr(p)}{Pr(q)}$$

Let's work through this with our rain example. As already stated, Pr(p) is 0.1. Suppose I believe that on nine out of ten days where it is raining at 12.00 noon it was overcast and cloudy at 9.00 a.m., so that Pr(q/p) is 0.9. Finally, suppose that I believe that it is cloudy and overcast at 9.00 a.m. on three out of every twenty days. Then Pr(q) will be 0.15. If we plug these figures into the equation, then the value of Pr(p/q) comes out at 0.6.

Now that we have an understanding of Bayes's theorem and its relevance to belief change in the light of new information, we can come to appreciate the application for Bayes's theorem to vision. Let's do this through an examination of the above case where a picture of a banana has been altered to make it grey yet, nevertheless, subjects perceive it to be yellow.

Rather than talking about propositions, it will be helpful to talk about hypotheses and evidence for the truth of a particular hypothesis. Suppose that an individual is presented with a picture of a grey banana. The wavelength of the light reflected onto the individual's retinas from the picture is one that is often associated with a conscious visual experience of grey. This retinal stimulation provides evidence as to the colour of the picture. However, this evidence is not unambiguous, as it could be produced by a picture of a grey banana in normal lighting conditions or a picture of a yellow banana in aberrant lighting conditions. Hence, the visual system has to work out which of these possible scenarios is the most likely. In other words, the visual system of the individual is faced with a choice between two competing hypotheses. According to the first, H1, the picture is grey in colour. According to the second, H2, it is yellow in colour.

Bayes's theorem can help in deciding which of these competing hypotheses is the most likely to be true. Utilizing Bayes's theorem involves calculating the value of the probability that H1 is true given the evidence (that is, the retinal stimulation) and the value of the probability that H2 is true given that evidence. If the former value is higher than the latter, then the visual system will prefer H1 to H2 given the evidence available to it. If, on the other hand, the latter value is higher than the former, it will prefer H2 to H1 given the evidence available to it. In other words, the visual system has to compute the following two equations:

$$1 \quad Pr(H1/E) = \frac{Pr(E/H1) \times Pr(H1)}{Pr(E)}$$

$$2 \quad Pr(H2/E) = \frac{Pr(E/H2) \times Pr(H2)}{Pr(E)}$$

In both cases the value for Pr(E) will be just the same, so what is crucial is which of Pr(E/H1) × Pr(H1) and Pr(E/H2) × Pr(H2) has the highest value. Now the value of Pr(E/H1) will be higher than that of Pr(E/H2). This is because the visual system would regard it as more likely that a picture of a grey banana would give rise to the retinal stimulation that constitutes E than would a picture of a yellow banana; the latter would be expected to cause a more 'yellowy' retinal stimulation. However, with respect to the values of $Pr(H1)$ and $Pr(H2)$, the situation is reversed. This is because of the visual system's prior knowledge that bananas are usually yellow and rarely grey. The size of this discrepancy swamps the size of the reverse discrepancy between the values of $Pr(E/H1)$ and $Pr(E/H2)$ so that the value of $Pr(E/H2) \times Pr(H2)$ comes out higher than that of $Pr(E/H1) \times Pr(H1)$. As a result of this, the value of equation 2 is greater than that of equation 1. Consequently, the visual system will prefer H2 to H1; given the evidence E, it will regard H2 as more likely to be true and will endorse it as the upshot of its Bayesian computations. The downstream consequence is that the individual will come to experience the pictured banana as being yellow.

What this particular example brings out is that the visual system's prior estimate of the probability of the competing hypotheses is of crucial importance. It is because the visual system placed such a high value on the probability of H2 relative to H1 that that hypothesis won out despite the fact that the evidence E fits better with H1 than it does with H2. A natural way of characterizing the situation is to say that the visual system had prior knowledge that bananas are usually yellow and very rarely grey. But this raises the question of where that prior knowledge came from. The two obvious candidates are either that this is an item of innate knowledge or that it was learned on the basis of encountering many yellow bananas and few, if any, grey bananas.

In sum, then, some rather surprising experimental data concerning colour experience can be readily accounted for by the view that the visual system operates by applying Bayes's theorem.

What philosophical issues do the data and theories that I have been describing in this section have? Two immediate issues arise that I will describe briefly without examining in great detail. The first relates to consciousness. Although much visual processing is unconscious, few cognitive scientists would deny that we do typically have conscious visual experiences as an upshot of visual processing. The relevant notion of consciousness here is phenomenal or qualitative consciousness; in other words, it has to do with the 'what it is like' aspect of visual perception (Nagel, 1974). For example, when I see a cricket ball as being red and spherical, there is 'something that it is like' to see it that way, and what it is like is different from what it is like to see an object as being cuboid and blue.

Now many philosophers and cognitive scientists hold that there is a close connection between the brain and consciousness in that activity in the brain gives rise to conscious experiences.[18] Some would want to go further and argue that there are likely to be neural correlates of consciousness, so that, for example, the experience of seeing red is always associated with the occurrence of a particular type of neural event, whereas the experience of seeing blue is always associated with some other specific event. To my mind, the question of whether or not there are neural correlates of consciousness is an empirical matter to which I maintain an open mind. However, some[19] want to go further and argue that, insofar as they are real, conscious phenomena reduce to neural phenomena and so can be explained in terms of them. To adopt such a view is to question the well-known claim of Chalmers (1996) that explaining consciousness is the hard problem in the philosophy of mind and that this hard problem cannot be solved by appealing to the brain.[20] The issues here are complex and highly contentious, and I will not explore them any further.

A second important issue relates to the metaphysical and epistemological status of our visual experiences of the world. Do we see the world as it really is independently of our experiences? Indeed, do we succeed in seeing a mind-independent reality at all as opposed to a mental construct? The prominent cognitive neuroscientist Chris Frith (2007) argues for a negative answer to these questions on the basis of his view that the brain is a 'Bayesian machine'. The basic idea is this: because how we see the world is shaped by our prior beliefs and expectations – as opposed to being exhaustively determined by how the world impinges on our sensory surfaces – what we perceive is a model existing in the mind rather than the external world itself. As Frith puts it:

> Our brains build models of the world and continuously modify these models on the basis of the signals that reach our senses. So, what we actually perceive are our brain's models of the world. They are not the world itself, but, for us, they are as good as. (2007: 134–5)

It is something of an understatement to say that this issue is complex and has a rich philosophical provenance. For example, Locke ([1689] 1975), though realist in overall perspective,[21] drew a distinction between primary and secondary qualities. Unlike primary qualities, secondary qualities are not objective features of reality but are projected onto the world by us. Locke placed colour on the secondary side of this divide. At the opposite extreme, Berkeley (1975) argued for the idealist view that the only things that exist are minds and mental entities. Therefore, 'physical objects' are collections of ideas. Another option is constituted by Kant's ([1781] 1998) transcendental idealism. For Kant, the world of our experience is a world of objects located in space and time that causally interact with one another. However, although there is a mind-independent reality, we do not experience this reality, as it is independent of us. Rather, our experience is shaped by *a priori* elements which the mind brings to experience. These *a priori* elements include space and time, which are the forms of intuition, and substance (that is, things) and cause, which are fundamental concepts (which Kant calls 'categories').

I mention these great thinkers not as a starting point to a discussion of Frith's claim but to emphasize that the issues are deep and complex and not to be taken lightly. I don't think that the emergence of Bayesian views of the brain move the debate on in any substantial way, and it is perfectly possible to square such a Bayesianism with any of the traditional episte- mological and metaphysical perspectives on the 'external' world and our relation to it. For example, while Frith says that we perceive our brain's models of the world, one could argue that one doesn't perceive these models but, rather, perceives the world through these models.

I now come to a third philosophical issue related to the views of visual perception discussed in this chapter, and this time I will discuss the issue in some detail. Accordingly, I will devote a whole section to it.

5 The location of perception

The discussion of visual perception in this chapter might be taken to indi- cate that vision is a phenomenon that takes place in the brain, and certainly that is the standard view in cognitive science. However, some philosophers and cognitive scientists have recently come to question this view and in doing so have drawn upon some of the findings that we have already exam- ined, particularly those relating to change and inattentional blindness. In rejecting the view that vision is a 'brain-bound' process, these thinkers have developed an alternative position, known as enactivism, that portrays vision as being intimately bound with bodily activity. Enactivism is related to the anti-representational views which I discussed in chapter 2. That is not to say that all enactivists deny that neurally realized representations are involved in vision; but, at the very least, they argue that such representations are not as central to vision as is traditionally thought within cognitive science.

Versions of enactivism have been championed by a number of theo- rists,[22] but my attention will focus on that developed by Alva Noë (2004) in his book *Action in Perception*, as this is arguably the clearest, best- developed and most influential example of the enactivist approach to vision. Noë doesn't go quite so far as to reject representations or the idea that processes internal to the brain play an important role in visual per- ception, but he makes a definite move in that direction. For central to his enactivist theory is an emphasis on the importance of bodily action to visual perception and the idea that sensation alone is not sufficient for fully fledged visual experience. Hence, visual perception cannot be under- stood as a process in the brain but is something done by a whole embodied animal embedded in an environment. Here is how Noë puts it:

> We ought to reject the idea – widespread in both philosophy and science – that perception is a process *in the brain* whereby the perceptual system

constructs an *internal representation* of the world. No doubt perception depends on what takes place in the brain, and very likely there are internal representations in the brain (e.g. content bearing internal states). What perception is, however, is not a process in the brain but a kind of skilful activity on the part of the animal as a whole. (2004: 2)

Noë contrasts his enactivism with a traditional computationalist approach, exemplified by Marr (1982), which he calls the 'snapshot conception' of perception (2004: 35), also dubbed the input–output theory by Hurley (1998). According to the snapshot conception, light reflected off surfaces in the external world is focused onto the retina, forming a two-dimensional retinal image. From this retinal image the job of the visual system is to construct a detailed representation of the distal scene. The retinal image is the sole source of input to the perceptual process, and the computational task of the visual system is very demanding as the retinal image is impoverished, providing little unambiguous information as to the outside world (Gregory, 1997). Nevertheless, the visual system manages to generate an output representation of the external world similar to a photograph, in that it is very detailed across its entirety.

Noë's initial complaint against the snapshot conception is that when we have a visual experience that experience isn't detailed across its whole range; rather, it is blurred and lacking detail around the edges. Moreover, the phenomena of change blindness and inattentional blindness further emphasize that when we have a visual experience we are typically unaware of much of the detail in the viewed scene. This suggests that the upshot of the visual process is not a detailed, photograph-like representation. In the light of this, Noë is motivated to compare vision to the activity of a blind person who perceives her environment by moving around it and probing it with her stick. The crucial point here is that this kind of perception involves ongoing activity, and it is something the blind person does rather than something that happens to her or within her. Similarly, argues Noë, vision involves bodily activity; in order to see the world we must use our bodies to move around and probe the world. Hence, being able to see involves having a practical skill that, by its very nature, involves utilizing the body. The upshot of this is that the relationship between visual perception and bodily activity is constitutive rather than merely causal. That is, it is not merely that how we act is causally influenced by how we see the world to be. Rather, to borrow an analogy from Clark (2006), the role of bodily activity in visual perception is akin to its role in dance: just as being able to dance requires having a body and dancing involves using that body, being able to see requires having a body and seeing involves using that body.

To make enactivism a viable alternative to the snapshot conception, more detail and argument is needed, and Noë attempts to provide this, as we will now see. He points out that there are sensorimotor contingencies – that is, disciplined relations between our bodily movements and how we sense the world when we make those movements. For example: as one moves closer to an object it will come to occupy a larger portion of one's visual field; if one moves one's head to the left then the position of a viewed object in one's visual field will shift to the right (and vice versa); and, as one changes one's position relative to a surface or object, its apparent shape changes; and so on. According to Noë, we have an implicit understanding of these sensorimotor contingencies that we exercise as we move around and visually perceive the world. This understanding is a practical skill rather than a body of propositional knowledge represented in the mind-brain, and without it we would have only sensations and not fully fledged visual experiences. This is because our sensations wouldn't have any significance for us, as their significance depends in a constitutive way on how we respond, or would respond, to them in moving around the world.

An example will help clarify matters. Suppose my eyes alight on a round saucer which, because of my spatial relation to it, looks elliptical in shape. Despite its looking elliptical, I also experience the saucer as being round, and this is an important aspect of the phenomenology of my experience.[23] In general, objects and surfaces appear in ways that do not line up with how we experience them to be, and this is something that it is incumbent on a theory of vision to explain. So, to return to the example of the saucer, the key question is this: how can the saucer look elliptical to me yet I experience it as being round? For Noë, the answer is not that I see the saucer as being elliptical but disregard my experience. Rather, there is a close link between looking one way from a particular perspective (for example, elliptical) and yet being objectively some other way (for example, round) that I fully appreciate. More specifically, I have an implicit (practical) understanding of how the look of something changes as one adopts different spatial perspectives with respect to it. Thus, although the saucer now looks elliptical, I appreciate that that is precisely how a round object would look from my current viewpoint. I exercise this knowledge as I move around the saucer so I can come to see that, despite its current look, it isn't elliptical but round. Hence, in virtue of my practical knowledge, my experience outstrips the current appearance of the objects I see.

Part of what Noë is getting at here is that, typically, you can't tell something's shape by viewing it from one position; rather, you have to move around it to see how its appearance changes. This is because different-shaped objects have different 'sensorimotor profiles', where 'the sensorimotor profile of an object is the way its appearance changes as you move with respect to it' (Noë, 2004: 78). Thus, a round object has one sensorimotor profile, an elliptical-shaped object another, and so on. Now suppose a particular object stimulates the retina in such a way as to look elliptical. It could be elliptical or it could be round. But someone armed with practical knowledge of the sensorimotor profiles of these distinct shapes could exercise that knowledge in moving around the object and observing how its appearance changes with such movements and so determine its objective shape. Of course, we sometimes experience an object as having a particular objective shape instantly without circumnavigating it. But even in such cases, argues Noë, experiencing the object in this way involves having expectations as to how its look would change were we to move around it, expectations that correspond to the relevant sensorimotor profile.

For Noë, the case of shape isn't an isolated example; rather, the phenomenon is endemic. For example, when I look at a tomato, I see not the whole object but only the surface that faces me. In fact, I don't see the whole surface in detail at any point in time but only part of the surface. Nevertheless, I experience the tomato as a whole object that has a behind that is hidden from my view, and I experience the surface as one that stretches beyond what I can clearly see. That I do so – something that Noë calls 'perceptual presence' (2004: 59) – is a product of the fact that I appreciate or expect that, were I to move behind the surface currently facing me, what I would then see would be a surface that looks a certain way (not unlike the way the surface currently exposed to me looks now). Once again, what is crucial is that I have a body that can move around the world and that I have implicit practical knowledge of how objects with different objective properties change the way they look as we move around them.

We now come to another important aspect of Noë's theory. This is his claim that the content of visual experience is virtual. The idea here is related to the phenomena of change and attentional blindness. Suppose I am at my computer desk. I have access to the Stanford Encyclopedia of Philosophy, to the whole of it, even though I haven't downloaded the entirety of its contents onto my desktop. The crucial point is that I could download any part of the Encyclopedia if I wished to do so. So the whole content is virtual in that respect. Now suppose I am viewing an object or surface. There won't be a detailed photograph-like representation in my mind generated from a retinal input. Rather, I will have only a dim and limited awareness of most of the object or surface that is before me. This explains change and inattentional blindness (I can be oblivious to substantial changes in the scene I am viewing and can fail to notice elements of the scene that I am not attending to). Nevertheless, I appreciate that I can access more detail if I so wish (this is the equivalent of downloading a particular article from an online encyclopedia). A key point is that, because any detail I might want (as to a particular element of an object or surface) is readily accessible if I utilize my sensorimotor skills, I don't need a detailed photograph-like representation in my brain. So, the ready accessibility of the world means that a considerable burden is taken off the mind-brain. As Noë puts it:

We have the impression that the world is represented in full detail in consciousness because, wherever we look, we encounter detail. All the detail
is present but it is only present *virtually*, for example, in the way that a
website's content is present on your desktop. (2004: 50)

It is here that Noë comes close to aligning himself with the views of Gibson
(1979) and Brooks (1991), as captured by the slogan that 'the world is its
own model', since, being readily accessible to us, we don't need a detailed
internal representation of the external world.

Should we be convinced by Noë's arguments? I think that he has done
a valuable service in drawing our attention to the dynamic nature of visual
perception, to the importance of acting in facilitating seeing, and to the significance of change and inattentional blindness. Nevertheless, my general
view is that most of these phenomena can be dealt with without abandoning the view that vision is a process of manipulating representations that is
firmly located in the brain. To see this, let's begin by considering his postulation of sensorimotor knowledge.

Noë thinks of our sensorimotor knowledge as being a kind of practical
knowledge or know-how. However, an option open to the advocate of the
traditional representational approach is to accept that we have sensorimotor knowledge but to view it as an instance of propositional knowledge
that is represented in the mind-brain. This would include propositional
knowledge of the sensorimotor profiles of different shapes that could be
drawn upon in perception in two ways. First, in working out how to act so
as to acquire further retinal stimulation to add to the retinal stimulation
already received, thereby expanding the evidence base as to the nature of
the viewed scene. The upshot of such computations will be motor instructions that cause relevant actions. Second, in working out the significance
of a series of visual representations derived from retinal stimulations gath-
ered over time as the individual explores the viewed scene.

To adopt this view would involve accepting that visual perception
typically involves a dynamic process unfolding over time in which bodily
activity plays a key role. But that is not self-defeating for two reasons. First,
bodily movement is valuable precisely because it results in the delivery of
new retinal stimulation that can be processed and then brought to bear
alongside previously processed retinal stimulation in an ongoing task of
determining the nature of the viewed scene. Second, bodily movement
is driven by representations in the form of motor instructions which
are themselves generated by means of processing representations. Thus,
although bodily movement plays a key role in facilitating perception, it is
not itself a part of the perceptual process but, rather, a tool that the visual
system utilizes to acquire the retinal stimulation that it needs to perform its
representational function.[24]

Noë objects to such a view of sensorimotor knowledge on the grounds that it is circular, as it postulates content involving representations in order to explain visual perception. In other words, in order to explain how we come to have perceptual experiences with particular contents (for example, one that represents an object as being round), it postulates complex representations featuring components with the target contents (for example, representations that represent how round objects look from various perspectives). I think that this objection is off-target for the following reason. The aim of the vision scientist is not to explain how representations get their content in non-semantic terms; that is, it is not to produce a naturalistic theory of content. Rather, it is to explain how output representations with a particular content are derived from input representations with some other content – for example, how representations representing the shape of distal objects are derived from representations representing patterns of retinal stimulation. Given this aim, there need be no loss of explanatory power if stored representations are postulated that have contents that overlap with those of the input and the output representations. To see this, consider an example. As a speaker of English, when I hear people utter sentences of that language I am generally able to work out the meaning of what they say. I am able to do this partly because I know the meaning of many words of English, and so represent each of those words as having a particular meaning. When I hear someone say 'the dog chased the postman', I am able to work out that they mean *the dog chased the postman*. To say that I do this by drawing upon my knowledge that the English word 'dog' means *dog* is hardly circular or lacking in explanatory power. By parity of reasoning, to say that my ability to work out that the saucer is round on the basis of its looking elliptical from various viewpoints relies upon stored knowledge of the relationship between apparent shape and actual shape is hardly to say something that is lacking in explanatory power.

My second objection to Noë's enactivism relates to its capacity to explain the phenomenology of visual experience. Recall that Noë claimed that we experience the objects that we see as having objective properties as well as appearance properties and that this is part of the phenomenology of vision and something that it is incumbent on a theory of vision to explain.[25] His attempted explanation appeals to sensorimotor knowledge. The problem with this is that it is far from clear how such knowledge, as it takes the form of knowledge-how or a body of expectations, could ground phenomenology or make the required phenomenological difference.[26] A champion of a representationalist perspective might argue that, in addition to generating representations that represent the appearance of objects from the viewer's perspective, the visual system generates representations as to the objective properties of those objects. To my mind, these latter representations are just as well suited as Noë's sensorimotor knowledge to grounding the phenomenology of our visual experiences, so his theory has no advantage on this front.

My third objection relates to the role of detailed representations in visual perception. As the quotation above indicated, Noë does not deny that there are representations in the brain; rather, his target is the snapshot conception that the endpoint of the visual process is a representation that represents the entirety of the viewed scene in rich detail. Now perhaps Marr held the snapshot conception in portraying vision as a bottom-up process whose sole input is a retinal image that is fully processed before any action is initiated. But, as we saw in the previous section, there have been significant developments in the representationalist framework post-Marr that find a place for top-down factors and fully engage with change and inattentional blindness. What these developments suggest is the following. The scenes that we confront are rich and complex, and to represent them in full detail is very demanding in processing terms given the impoverished and ambiguous nature of retinal stimulation. In order to reduce processing burdens on the visual system, which elements of the retinal input are pro-cessed is influenced in a top-down way by the subject's aims and interests. Thus, for example, because a subject in the Simons and Chabris (1999) experiment is concerned with counting the number of passes made by the players wearing white, her visual system doesn't fully process those aspects of retinal stimulation that do not relate directly to those players. Such details are, in effect, deemed irrelevant to the task at hand. It is important to note that this doesn't imply that detailed representations are not involved in vision. After all, the subject does extract significant detail relating to the passing activity of the players wearing white. Thus, there is a distinction between the claim that the visual system produces detailed representations relating to all aspects of the viewed scene and the claim that it produces detailed representations relating only to specific aspects of the viewed scene. In effect, by endorsing the second of these two claims, one can avoid Noë's enactivism while engaging with the phenomena of change and inat-tentional blindness.[27]

A fourth objection relates to Noë's claim that visual content is virtual. In order to read a particular article from the Stanford Encyclopedia of Philosophy I have to download it onto my desktop. When I do this the article is right before me and no longer (merely) virtual. What makes the whole Encyclopedia (or any part of it) virtual is that I have the capacity to download it and so make it present to me. Therefore, virtuality is potential presence. Applied to vision, this suggests that, when I look at a scene and am not aware of the rich detail of a particular element of it, I nevertheless have the capacity to become aware of that element. Within a traditional representationalist framework, becoming aware of that element would involve executing further visual processing so as to construct a detailed representation of it. In other words, the virtual nature of visual content, such as it is, is a product of the need of the visual system to deal with the potentially overwhelming demands of visual processing by being selective and in no way suggests that detailed representations are not central to visual experience.

In sum, then, Noë hasn't provided compelling reasons for endorsing enactivism. An updated theory of visual perception as a process of manipulating representations that is firmly located in the brain can sit happily with the phenomena of change and inattentional blindness and recognize that vision dynamically uses the body as a tool.

6 Conclusion

In this chapter I have explained how a deepening understanding of the workings of the brain and an associated development of brain-imaging technology have resulted in neuroscience coming to have a much more prominent role in cognitive science over the last two decades. This has been the case particularly with respect to the study of visual perception, where some surprising results have been obtained suggesting that vision is neither a single process nor a bottom-up one driven exclusively by retinal stimulation. Despite this general trend, there are some philosophers and cognitive scientists who have attempted to resist such a brain-centric approach to vision, arguing that vision should be seen as a bodily activity or skill. I have argued against this enactivist movement.

Conclusion

Over the previous six chapters I have provided a general survey of some of the major issues in the philosophy of cognitive science. These issues have included the following: the nature and role of representations in cognition; the extent to which we reason rationally; the nature of the concepts we use in categorizing objects and reasoning and their relationship to perceptual states; the organization of the mind and the role of evolution in sculpting that organization; the basis of our mastery of language and the means by which we acquire that mastery; the relationship between the brain and cognition; and the nature of visual processing and its relation to action.

In chapter 1, I made the point that cognitive science emerged from a commonsense conception of humans but that we shouldn't expect it to cling onto or ultimately vindicate that conception. This is because sciences often undermine their folk ancestors (think how much quantum physics and relativity theory diverge from folk physics). That raises the question of the extent to which cognitive science has undermined the commonsense conception we have of ourselves.

One implication of the narrative of this book is that much of common-sense remains unscathed. For an ordinary person would be able to relate to many of the issues I have discussed – for example, questions about how we reason, categorize objects, develop language, perceive the outer world, and so on. This suggests that cognitive science has not thrown away our commonsense conception of ourselves. However, in attempting to answer these questions, cognitive scientists often make claims that are surprising or even bizarre from the commonsense perspective. For example, that much of what goes on in the mind is either unconscious or subpersonal; that we routinely fail to reason rationally; that the mind is a computer or a connectionist network; that action is constitutively linked to cognition; that cognitive processes stretch beyond the skull; that the mind is composed of distinct task-specific components grounded in our distant evolutionary past; that from an early age children operate with quasi-scientific theories which develop in a way that reflects the history of science; that languages have universal features that are grounded in the innate structure of the mind; that visual perception decomposes into distinct processes for guiding action and categorizing objects; and so on. It is this general sense that cognitive science is concerned with us humans as we see ourselves but also delivers very surprising theories about the basis of our distinctive mentality that makes the subject so interesting and exciting.

Throughout this book I have described many controversies, some of which go back to the early days of cognitive science and are still very much alive today. The existence of such controversies indicates that cognitive science (and its philosophy) is not a discipline where widespread agreement reigns. Rather, there is a pluralism of approaches and points of view. Moreover, much of the disagreement relates to fundamental matters. This shouldn't be all that surprising or disheartening for two reasons. First, cognitive science is a relatively young science and so it is a little too early for it to have conquered its subject matter. Second, in engaging in cognitive science we are trying to use our cognitive capacities to understand our cognitive capacities, and there is a real possibility that those capacities are just not up to the job. After all, one wouldn't expect a rat to be able to come to understand the cognitive workings of rats, even though there are many things that rats can understand. I'm not suggesting scepticism here about the ultimate prospects for cognitive science but merely drawing attention to the point that we should expect cognitive science to be a tough and long haul.

I have tried to be fair and even-handed in discussing the controversies that enliven the philosophy of cognitive science. However, my approach has not been of the 'Professor X says this and Professor Y says that' variety. Rather, I have expressed my own views and presented arguments for them. I make no apologies for this, as I see it as being fundamental to stimulating the reader to engage their own philosophical powers and reach their own conclusions.

My general perspective is one of being sympathetic to classical computationalism, the modularity thesis (especially with respect to language), the idea that domain-general learning mechanisms do not underpin all key aspects of cognitive development, and the view that cognition is constitutively distinct from (but causally related to) action and very much located in the brain. In the eyes of some, this might make me fusty, old-fashioned and somewhat conservative, especially in the light of the four Es (the view that cognition is embedded, embodied, extended and enactive). However, I baulk at this accusation for several reasons. First, though it is true that my general sympathies lie with views that have been prominent since the early days of cognitive science, the views of those who champion an alternative vision have an equally long-standing heritage and have hardly appeared from nowhere in the last decade. For example, friends of the four Es have drawn inspiration from Heidegger, Merleau-Ponty, Wittgenstein, J. J. Gibson and other figures of notable vintage.

Second, in an important sense, what makes a view radical is not so much how long it has been around or how widely endorsed it has been in the history of an academic discipline. After all, it would be odd to deny that quantum physics and relativity theory were radical merely because they have been widely endorsed features of the intellectual landscape for nearly a century. Rather, whether or not an idea is radical has to do with how much it transforms our commonsense view of the subject matter in question. On this front, arguing that cognition is based on neurally realized computers or that there is a language faculty that undercuts the role of domain-general learning mechanisms in language development surely counts as radical.

Third, the views with which I have expressed sympathy are not set in aspic; rather, they are dynamic and have developed over the years to deal with fresh data and the emergence of new research tools. A good example here relates to visual perception, where a classical computationalist is hardly obliged to cling onto the letter of Marr's theory of vision but can recognize the importance of top-down factors in visual processing, draw upon data gleaned from neural imaging, accept that bodily action plays an important ongoing role in facilitating vision, and so on.

Some of the debates I have discussed have a history of several decades, whereas others have emerged only recently. All of these debates are interesting and important, but in academic disciplines fashions change and theorists move away from once hot topics when they feel they have little more to say or that the available ideas have been exhausted. Given this, it might be helpful to identify what I regard as cutting-edge topics at this point in time, the issues that are likely to dominate the philosophy of cognitive science for the next few years.

With respect to the nature and role of representations in cognition, I regard the cutting-edge topics as these: What is the role of biases and heuristics in cognition and to what extent are we rational in our thinking? Are there distinct systems involved in personal-level reasoning? Is the thinking associated with language distinct from non-linguistic thought? Can successful non-representationalist and connectionist models of simple systems and basic cognitive capacities be scaled to offer plausible explanations of high-level human cognition? And to what extent do our actual cognitive processes extend beyond the skull?

With respect to modularity, cutting-edge topics include the following: What is the role of top-down factors and attention in perception? Does the use of heuristics make high-level reasoning computationally tractable? To what extent do evolutionary considerations tell in favour of massive modularity? How does social scaffolding aid central cognition? And what is the role of domain-specific theories in cognitive development and to what extent are they supplemented by domain-general learning?

With respect to concepts, among cutting-edge topics are these: What role do perceptual representations play in our concepts? Can perceptually based theories of concepts account for abstract concepts? Are philosophers and psychologists engaged in incommensurable projects when they theorize about what they call 'concepts'? To what extent do we share concepts? What is the relationship between conceptual development in children and scientific development? What is the relationship between concepts and language? And to what extent are we committed to essentialism and what is the significance of that commitment?

With respect to language, cutting-edge topics include the following: Are there any significant linguistic universals? What is the role of domain-general statistical learning mechanisms in language acquisition? Assuming that language is specific to humans, why is this the case? And to what extent are a child's linguistic experiences impoverished?

With respect to the brain, cutting-edge topics include these: To what extent is the brain a Bayesian machine? How plastic is the brain? Do cognitive phenomena reduce to neural phenomena? Do cognitive processes such as perception take place in the brain or are they constitutively linked to bodily action? And what is the relationship between phenomenal consciousness and the brain?

I don't expect agreement as to how these questions should be answered to emerge any time soon. However, the debate as it unfolds over the coming years is sure to be fascinating and of fundamental importance if we are ever to get to the heart of the basis of our humanity.

Notes

CHAPTER 1 COGNITIVE SCIENCE AND THE PHILOSOPHY OF COGNITIVE SCIENCE

1 This is known as the principle of compositionality and is associated particularly with Gottlob Frege ([1914] 1979).
2 Richard Gregory (1966) offers the classic expression of this view.
3 In arguing along these lines he echoes views of Chomsky (2000) and Fodor (1987).
4 For example, that constructed by Mendel as a result of his work breeding peas.
5 This is sometimes called the Quine–Duhem thesis or the thesis that theory is under-determined by observational data.
6 Such a view is championed by Keil (1989), Bloom (2004) and Carey (2009).
7 Wilhelm Wundt and William James are the key figures, being responsible for, respectively, the establishment of the first psychology laboratory in Europe and the first psychology laboratory in the United States.
8 The term 'representation' was Kant's ([1781] 1998) preferred term; other philosophers used terms such as 'idea' and 'impression'.
9 One that is particularly associated with Hume.
10 Indeed, Dennett (1978a) argues that the objection can be found in the work of Hume.
11 Kenny (1984), Baker and Hacker (1984), Dummett (1993) and McDowell (1994) are philosophers influenced by Wittgenstein who have developed similar objections against the invocation of representations.
12 Such a mechanical procedure is known as an algorithm.
13 The most prominent champions of this theory were U.T. Place (1956), Herbert Feigl (1958) and J.J.C. Smart (1959).
14 This is known as generative linguistics.

CHAPTER 2 REPRESENTATION AND COMPUTATION

1 Key classical computationalist works include Fodor (1975), Newell and Simon (1976), Marr (1982) and Pylyshyn (1984).
2 In particular, in Rosenblatt's (1962) work on perceptrons, whose short-term influence was undermined by Minsky and Papert's (1969) critique.
3 This mechanistic perspective is captured in Boden's (2006) characterization of cognitive science as being based on the idea that the mind is a machine. Also see Crane (2003).
4 For further details, see Carnie (2013).
5 This view is widely associated with Frege ([1914] 1979).
6 Pylyshyn (1984) calls this mapping the instantiation function.
7 Indeed, in principle, how the symbols of a language are realized in a given machine could change over time.
8 Of course this doesn't exhaust the possibilities for the classical computationalist, for

she could take the view that some subpersonal phenomena are computational while others are not, or that some personal-level phenomena are computational while others are not.

9 These stages are not to be confused with his three levels described above. For each of these stages could be characterized at each of Marr's three levels of explanation.

10 See Cain (2002: ch. 2) for a more detailed account of Marr's theory.

11 John Searle has provided some of the best-known objections to classical computationalism. In the Chinese Room argument (Searle, 1980) he objects that no computer, however it is programmed, is capable of such an intelligent phenomenon as understanding. More recently, he has argued that syntactic properties are not objective features of the world and so cannot appear in legitimate scientific explanations (1992). I have discussed these objections elsewhere (Cain, 2002) and so won't consider them any further here. Neither will I consider the objections to classical computationalism provided by Daniel Dennett (1978c, 1987, 1991a) in developing his instrumentalist view of cognition. For my views on Dennett's instrumentalism see, once again, Cain (2002).

12 Patricia Churchland (1986) also develops a version of this argument.

13 Also see Gallistel and King (2009).

14 Newell and Simon's (1961) General Problem Solver is the stand-out example of this.

15 Prominent examples are Evans (1984, 1989), Sloman (1996) and Stanovich (1999).

16 After all, as Grice (1975) emphasized, successful communication involves not merely understanding the literal meaning of what is said. In addition it requires appreciating the conversational implicature of what is said. If this is right, then we shouldn't be surprised if the subjects in the experiment understood 'Linda is a bank teller' as implying that Linda is a stereotypical bank teller with the associated political commitments, especially as some of the other statements overtly refer to her political commitments.

17 'Intentionality' is a technical term that means roughly 'aboutness'. A symbol has intentionality because it is about a particular thing or type of thing and represents what it is about as being a particular way.

18 Daniel Dennett (1987, 2013) rejects this idea that our mental states have original intentionality. For him all intentionality is derived and a matter of interpretation. Donald Davidson (1980) is another prominent advocate of interpretationalism, though his work does not involve any substantial engagement with cognitive science.

19 For an overview, see Cain (2002: ch. 5).

20 Burge (2010) and Chalmers (2010) are two prominent philosophers who reject the assumption that the meanings of our mental representations are grounded in more basic physical properties in terms of which they need to be explained.

21 Which are often called neural networks because of their brain-like nature.

22 For an extensive overview, see Bechtel and Abrahamsen (2002).

23 Within a classical computationalist perspective such as that developed by Fodor, concepts are symbols of LOT.

24 Which, recall, is made up of eighty units.

25 One of the most prominent champions of simple recurrent networks has been Jeffrey Elman (1991). In particular, he has attempted to design such networks in order to model language acquisition.

26 The input–output behaviour of such gates corresponds to the truth tables used to define such logical symbols as AND and OR in elementary logic.

27 Fodor and Pylyshyn (1988) argue along these lines.

28 This point is most forcibly made with respect to language development, where

Chomsky (1986) argues that the linguistic information that children receive in their early years is impoverished. See chapter 5 for further discussion.

29 For example, Varela et al. (1991), Thelen and Smith (1994), Beer (1995) and van Gelder (1995). More recently, the anti-representationalist approach has been championed by Chemero (2009) and Hutto and Myin (2013).

30 Hurley (1998) evocatively characterizes this perspective as viewing perception and cognition as being like the filling of a sandwich.

31 SHAKEY was a robot developed at Stanford Research Institute's Artificial Intelligence Center in the 1970s.

32 Keijzer (1999) develops an anti-representationalist approach consonant with those of van Gelder and Brooks that discusses animals, ranging from insects such as the Sphex wasp to lions.

33 Underlying this difference are differences in brain anatomy. Humans have a much larger and more developed frontal cortex than any other species. The frontal cortex is standardly identified as the seat of higher-level thought and planning. See chapter 6 for further discussion.

34 Clark (2008, 2014) characterizes the processes of reflecting on counterfactual, imaginary and distal situations as being representation hungry.

35 For sympathetic perspectives, see Clark (2008), Rowlands (2003) and Menary (2007). For a critical perspective, see Adams and Aizawa (2008), Rupert (2009) and Shapiro (2011). Menary (2010) is a helpful collection.

36 This was the central point of George Miller's (1956) landmark work on memory.

CHAPTER 3 MODULARITY

1 Or language faculty or organ, as Chomsky (1980, 1986) prefers to call it.

2 Arguably, this is a misrepresentation of Chomsky's position. See Collins (2004) for discussion.

3 Some theorists – for example, Carruthers (2004) – argue against understanding domain specificity as involving being sensitive to only a narrow range of inputs. For them, being domain specific relates to function – for example, having a particular subject matter or a particular problem to solve.

4 For example, by Carruthers (2006a) and Sperber (1996).

5 See Frith (2007: 48–50) for details.

6 Frith (2007) and Smith and Kosslyn (2007) are good examples.

7 Samuels (2006) makes a similar point.

8 Cowie (2008) cites Clahsen and Almazan (1998), Bello et al. (2004) and Bellugi and Lai (1998) as arguing for a similar view with respect to the linguistic capacities of WS subjects.

9 Also see Smith et al. (2010).

10 Accessible accounts of theta theory can be found in Adger (2003), Carnie (2013) and Radford (2009).

11 Louise Antony (2003) provides a nice example of this where astronomers date supernova on the basis of pictures on centuries-old Native American pottery.

12 The key figures here include Tooby and Cosmides (1992), Sperber (1996), Pinker (1997) and Carruthers (2006a).

13 The key advocates of this view of psychological explanation are Fodor (1968a), Dennett (1978a) and Cummins (1983).

14 Spelke (2003) also postulates modular core systems for numerosity and geometry.

15 See Prinz (2012) for a critique of such nativism.

16 Indeed, Gopnik (2003) applies such an approach to language acquisition, thereby rejecting the existence of a language module.

17 These include Scholl and Leslie (1999), Baron-Cohen (1995), Segal (1996) and Carruthers (2006a).

18 That is, linguists working in the Chomskyan generative tradition.

19 See chapter 5 for more on the topic of language and its modular status.

CHAPTER 4 CONCEPTS

1 See Margolis and Laurence (2007) for discussion and critical evaluation.

2 See Elbourne (2011), Murphy (2002), Bloom (2000) and Jackendoff (2012) for a defence of this perspective on the relationship between concepts and word meaning.

3 For example, Susan Carey (1985, 2009) argues that the content of a child's concepts changes as they develop towards adulthood.

4 Jesse Prinz (2002) agrees and labels this requirement for a theory of concepts to explain how it is possible for distinct people to share their concepts the publicity constraint.

5 This is an objection that Berkeley (1975) famously directed at Locke.

6 Though probably not one that Wittgenstein would have endorsed.

7 A lexical concept is a concept expressed by a morphologically simple word of a language. Both DOG and GAME are examples of lexical concepts, in contrast to the concepts THE QUEEN'S FAVOURITE GAME and THE DOG THAT LIVES NEXT DOOR.

8 Another prominent philosopher committed to the same conclusion about the indefinability of concepts is W. V. O. Quine (1951), in virtue of his rejection of the analytic–synthetic distinction.

9 See Rosch (1978), Rips et al. (1973) and Hampton (1979) for prominent defences of this view.

10 See Medin and Shaffer (1978) for a prominent expression of the exemplar theory.

11 Perhaps a qualification is needed here: one can have a feature that has its own prototype so long as that feature is inserted on the basis of identifying something external as satisfying the prototype in question.

12 Prominent advocates of the theory theory include Gelman (2003), Carey (2009), Spelke (2003) and Gopnik and Meltzoff (1997).

13 Gopnik and Meltzoff (1997) argue for such a view.

14 For example, Keil (1989) and Gelman (2003).

15 This is the so-called disjunction problem.

16 Recall that a lexical concept is one that is expressed by a morphologically simple word: DOG and COW are lexical concepts, but ANGRY BROWN COW is not.

17 See McGinn (1989a), Rowlands (2003) and Wilson (2004) for representative examples.

18 Not everyone has been convinced by Putnam's argument and its extension. Prominent critics include Chomsky (2000), Searle (1983), Crane (1991) and Farkas (2008).

19 See Cain (2013) for an extended development of this argument.

CHAPTER 5 LANGUAGE

1 For example, Donald Davidson (1975).

2 For helpful accounts of the development of Chomsky's views over the years, see 'The

Editors' Introduction' to Chomsky (2002), as well as Collins (2008), Boeckx (2010) and McGilvray (2005, 2013).

3 Recent advocates of this commonsense view from within these disciplines include Devitt (2006), Everett (2012) and Tomasello (2003).

4 Michael Dummett (1993: ch. 3), one prominent advocate of the commonsensical view, regards this knowledge as a practical skill rather than an internal representation.

5 See chapter 3 for general discussion of the modularity thesis.

6 The idea that there is this intermediate level of structure is the core element of X-bar syntax (Chomsky, 1970; Jackendoff, 1977). Hence, a constituent (be it a tense item, a noun, a verb, a preposition, or whatever) can head both an X′ structure (for example, a T′, an N′, a V′, a P′, or whatever) and a full phrase (for example, a TP, an NP, a VP, a PP, or whatever).

7 What we pre-theoretically call sentences are complementizer phrases.

8 A clause is a structure consisting of a subject and a predicate.

9 For illumination I recommend consulting Radford (2009), Carnie (2013), Larson (2010) or Adger (2003).

10 See Laurence and Margolis (2001) for a helpful discussion of the poverty of the stimulus argument.

11 The asterisk here indicates that the structure is unacceptable to most speakers of English.

12 A principle is a rule of great generality. Chomsky and his followers now tend to talk of principles rather than rules, as the term 'rule' tends to carry with it the implication of narrowness and specificity.

13 Think how different social groups have independently invented similar tools, such as axes, knives, and bows and arrows, in order to deal with a problem they all face.

14 When saying what somebody thinks, the Pirahã use the verb 'to speak'. In effect, they put words into the mouths of other people to specify what they think.

15 This point comes out clearly in Tomasello's (2008) reflections on communication.

16 Such as those made in the following claims:
All languages employ recursion.
All languages build structures by means of a binary merging operation.
In all languages syntactic operations – such as movement agreement and case assignment – are local.

17 To recap, a child would receive negative data were she explicitly told that a particular ungrammatical structure was ungrammatical. Chomsky's (1972) claim is that children are rarely presented with such data.

18 An intransitive verb is a verb that cannot take an object. Examples include 'run', 'fall' and 'laugh'. A causative is a verb that expresses a causal relationship, for example, 'melted' in 'the sun melted the snowman'.

19 Helpful books on language acquisition that discuss the conflict between Chomsky and Tomasello's competing approaches include Russell (2004), Ambridge and Lieven (2011) and Rowland (2014). Eve Clark (2009) develops a perspective on language acquisition that is quite similar to that of Tomasello, as does Vyvyan Evans (2015).

20 As pattern-finding involves uncovering statistical regularities in input data, in finding a place for pattern-finding in language acquisition, Tomasello is aligning himself to a growing number of scholars who argue that languages are acquired by means of domain-general statistical learning mechanisms. The classic work here is by Saffran,

Aslin and Newport (1996), who argue that infants as young as eight months can extract three-syllable words from continuous speech in an artificial language after only two minutes of exposure to that language. For an enthusiastic overview of the statistical learning approach to language acquisition, see Prinz (2012: ch. 6).

21 Many philosophers and psychologists use the term Theory of Mind (ToM for short) when talking about mind-reading. For a general overview of the role of mind-reading in everyday life, see Bloom (2004) and Epley (2014).

22 This is a case of what Tomasello (2008) calls pantomiming.

23 For example, Grice (1957), Lewis (1969, 1975), Gilbert (1989), Millikan (1998, 2003), Dummett (1993) and Devitt (2006).

24 This is particularly true of connectionist approaches, such as that of Rumelhart and McClelland (1986) on past tense formation and Elman's (1991) work on long-distance agreement.

25 Detailed versions of construction grammar have been developed by a number of scholars, including Goldberg (1995, 2006), Croft (2001), Lakoff (1987) and Langacker (1987, 1991).

26 The test was first carried out by the psychologists Wimmer and Perner (1983), but the basic idea is due to the philosopher Daniel Dennett (1978d).

27 Perner (1991) goes so far as to suggest that a failure to pass the false-belief test shows that children under the age of four do not have a concept of belief, since to have such a concept one has to appreciate that it is possible for a belief to be false.

CHAPTER 6 THE BRAIN AND COGNITION

1 This point about the reductionist perspective and the implication that explanatory power lies at the lowest level is clearly made by Patricia Churchland (1986, 2013).

2 For example, Paul Churchland (2005).

3 The Spanish neuro-anatomist Santiago Ramon y Cajal is usually credited with the discovery that the brain is made up of cells. He was awarded a Nobel Prize for physiology and medicine for his work in 1906 jointly with Camillo Golgi, who, rather ironically, opposed the cellular theory despite developing the staining technique that Cajal employed to motivate his theory.

4 Glial cells are kinds of cells that far outnumber neurons in the brain. They have traditionally been seen as performing 'housekeeping' tasks which enable the neurons to keep functioning, but recent research suggests that they play a more direct role in the development and functioning of the brain and in neurological disorders (Barres, 2008).

5 Thereby echoing a typical contemporary digital computer where representations or symbols are encoded by means of strings of 1s and 0s that are themselves constituted by electrical states of the computer's internal circuits.

6 I'm not going to discuss the main competing theories of content here. For an overview, see Cain (2002: ch. 5) and Braddon-Mitchell and Jackson (2007: ch. 11).

7 The hierarchical organization of the brain from sub-cellular structures upwards is very effectively described by Adam Zeman (2008).

8 The presence of Wernicke's area in only one of the temporal lobes is an example of the partial asymmetry of function between the two hemispheres that I mentioned above.

9 The ventricles don't play a direct role in cognition. One of their key functions is to cushion the brain against impact and sudden movement.

10 For more details on the structure and workings of the brain, I recommend Kandel et al. (2012) and Squire et al. (2013).

11 Cognitive scientists and medics who work with such patients tend to emphasize their incredible luck in having the patients referred to them in the first place and the serendipitous aspect of their appreciating the specific nature and potential significance of their cognitive deficits. See Goodale and Milner (2004) and Zeman (2008) for examples of such statements.

12 For a helpful examination of the strengths and limitations of cognitive neuropsychology, see Stone and Davies (2012) and Davies (2010).

13 Marr's (1982) theory is an example of a bottom-up theory of vision.

14 This and many other related experiments are discussed by Chabris and Simons (2010) in an entertaining book-length study.

15 Conducted by Levin et al. (2002).

16 Examples are Frith (2007), Clark (2013) and Hohwy (2013).

17 Objective probability, in contrast, relates to objective frequencies that hold independently of anyone's expectations. See Papineau (2012) for a clear account of the distinction between subjective and objective probability.

18 Perhaps Daniel Dennett (1991b) is an exception, as he wants to reject the widespread conception of consciousness as an inner experience that has a 'what it is like' character.

19 For example, Francis Crick (1994), Christof Koch (2004), Paul Churchland (2007) and Patricia Churchland (2013).

20 For other prominent expressions of scepticism regarding the explanatory power of appeals to the brain with respect to consciousness, see Saul Kripke (1980), Frank Jackson (1986) and Joseph Levine (2001).

21 Where realism is the view that there is a mind-independent reality and that we have the capacity to acquire knowledge of that mind-independent reality.

22 For example, Hurley (1998), O'Regan and Noë (2001), Thompson (2007), Chemero (2009) and Hutto and Myin (2013). Hutto and Myin (2013: ch. 2) provide a helpful account of the relations between the different versions of enactivism, including Noë's sensorimotor enactivism, Thompson's (2007) autopoietic enactivism and their own radical enactivism.

23 Noë's emphasis on the phenomenology of vision is a potential reason for resisting Block's (2005) charge that he is a behaviourist.

24 It should be pointed out that it is possible to view neurally realized representations as being central to perception *and* as having a tighter connection to action than I have suggested. For example, one might follow Evans (1985) in arguing that the content of a perceptual representation is a matter of the dispositions to action that it grounds. Alternatively, one might endorse Bence Nanay's postulation of pragmatic representations. These are representations 'that represent perceived objects in an action-orientated manner' (2013: 4).

25 Jesse Prinz (2006b) questions whether this feature of our experiences is part of their phenomenology.

26 Pierre Jacob and Andy Clark both raise the same worry. Jacob writes: 'in spite of Noë's explicit disclaimer . . . his enactive account does seem to turn the experience of perceptual constancy into the conclusion of a plain inferential process' (2006: 8). Clark writes: '[t]he obvious drawback with this proposal is that it leaves unexplained why knowledge concerning the relevant sensorimotor space should result in the experience of anything at all' (2006: 3).

27 Jacob (2006) has a somewhat different argument for the claim that change blindness

doesn't in itself vindicate enactivism. Drawing upon Dretske (2004), he points out that, in order to notice changes in a viewed scene, one needs to be able to compare the scene as it was at different points in time. Several distinct factors could explain a failure to make that comparison even though a detailed representation of the scene at each point of time was generated by the visual system. For example, the representation of the scene at the earlier point in time might not reside in memory for a sufficient amount of time to allow a comparison with the representation of the scene at the later point in time. Prinz (2006b) makes a very similar point.

References

Adams, F., and Aizawa, K. (2008) *The Bounds of Cognition.* Oxford: Blackwell.

Adger, D. (2003) *Core Syntax: A Minimalist Approach.* Oxford: Oxford University Press.

Ambridge, B., and Lieven, E. V. M. (2011) *Child Language Acquisition: Contrasting Theoretical Approaches.* Cambridge: Cambridge University Press.

Antony, L. (2003) 'Rabbit-Pots and Supernovas: On the Relevance of Psychological Data to Linguistic Theory', in A. Barber (ed.), *Epistemology of Language.* Oxford: Oxford University Press.

Baillargeon, R., Scott, R. M., and He, Z. (2010) 'False-Belief Understanding in Infants', *Trends in Cognitive Sciences* 14: 110–18.

Baker, G. P., and Hacker, P. M. S. (1984) *Language, Sense and Nonsense: A Critical Investigation into Modern Theories of Language.* Oxford: Blackwell.

Baron-Cohen, S. (1995) *Mindblindness: An Essay on Autism and Theory of Mind.* Cambridge, MA: MIT Press.

Barres, B. B. (2008) 'The Mystery and Magic of Glia: A Perspective on their Roles in Health and Disease', *Neuron* 60(3): 430–40.

Barsalou, L. W. (1999) 'Perceptual Symbol Systems', *Behavioral and Brain Sciences* 22: 577–660.

Bayne, T. (2010) *The Unity of Consciousness.* Oxford: Oxford University Press.

Bayne, T. (2013) *Thought: A Very Short Introduction.* Oxford: Oxford University Press.

Bechtel, W., and Abrahamsen, A. (2002) *Connectionism and the Mind: Parallel Processing, Dynamics, and Evolution in Networks.* 2nd edn, Oxford: Blackwell.

Beer, R. (1995) 'A Dynamical Systems Perspective on Agent–Environment Interaction', *Artificial Intelligence* 72: 173–215.

Bello, A., Capirci, O., and Volterra, V. (2004) 'Lexical Production in Children with Williams Syndrome: Spontaneous Use of Gesture in a Naming Task', *Neuropsychologia* 42(2): 201–13.

Bellugi, U., and Lai, Z. (1998) 'Neuropathological and Cognitive Alterations in Williams Syndrome and Down Syndrome', *Faseb Journal* 12(4): A355.

Berkeley, G. (1975) *Philosophical Works: Including the Works on Vision,* ed. M. Ayers. London: Dent.

Bermúdez, J. L. (2010) *Cognitive Science: An Introduction to the Science of the Mind.* Cambridge: Cambridge University Press.

Bickle, J. (2003) *Philosophy and Neuroscience: A Ruthlessly Reductive Account.* Norwell, MA: Kluwer Academic Press.

Block, N. (1986) 'Advertisement for a Semantics for Psychology', in P. A. French, T. E. Uehling and H. K. Wettstein (eds), *Studies in the Philosophy of Mind.* Minneapolis: University of Minnesota Press.

Block, N. (2005) 'Review of Alva Noë *Action in Perception*', *Journal of Philosophy* 102(5): 259–72.

Bloom, P. (2000) *How Children Learn the Meaning of Words*. Cambridge, MA: MIT Press.

Bloom, P. (2004) *Descartes' Baby: How Child Development Explains What Makes Us Human*. London: William Heinemann.

Bloomfield, L. (1933) *Language*. London: George Allen & Unwin.

Boden, M. (2006) *Mind as Machine: A History of Cognitive Science*. Oxford: Oxford University Press.

Boeckx, C. (2006) *Linguistic Minimalism: Origins, Concepts, Methods, and Aims*. Oxford: Oxford University Press.

Boeckx, C. (2010) *Language in Cognition*. Chichester: Wiley-Blackwell.

Braddon-Mitchell, D., and Jackson, F. (2007) *Philosophy of Mind and Cognition: An Introduction*. 2nd edn, Oxford: Blackwell.

Braine, M. D. S. (1976) *Children's First Word Combinations*. Chicago: University of Chicago Press.

Brooks, R. (1991) 'Intelligence without Representation', *Artificial Intelligence* 47: 139–59.

Brown, R. (1973) *A First Language: The Early Stages*. Cambridge, MA: Harvard University Press.

Bruce, V., and Young, A. W. (1986) 'Understanding Face Recognition', *British Journal of Psychology* 77: 305–27.

Burge, T. (2010) *The Origins of Objectivity*. Oxford: Oxford University Press.

Cain, M. J. (2002) *Fodor: Language, Mind and Philosophy*. Cambridge: Polity.

Cain, M. J. (2013) 'Concept Learning and Psychological Essentialism', *Review of Philosophy and Psychology* 4: 577–98.

Campos, J., and Stenberg, C. (1981) 'Perception, Appraisal and Emotion: The Onset of Social Referencing', in M. E. Lamb and L. R. Sherrod (eds), *Infant Social Cognition*. Hillsdale, NJ: Lawrence Erlbaum.

Carey, S. (1985) *Conceptual Change in Childhood*. Cambridge, MA: MIT Press.

Carey, S. (2009) *The Origin of Concepts*. New York: Oxford University Press.

Carnie, A. (2013) *Syntax: A Generative Introduction*. 3rd edn, Chichester: Wiley-Blackwell.

Carruthers, P. (2004) 'Practical Reasoning in a Modular Mind', *Mind and Language* 19: 259–78.

Carruthers, P. (2006a) 'The Case for Massive Modular Models of Mind', in R. A. Stainton (ed.), *Contemporary Debates in Cognitive Science*. Oxford: Blackwell.

Carruthers, P. (2006b) *The Architecture of the Mind: Massive Modularity and the Flexibility of Thought*. Oxford: Oxford University Press.

Chabris, C., and Simons, D. (2010) *The Invisible Gorilla and Other Ways our Intuition Deceives Us*. London: Harper.

Chalmers, D. (1996) *The Conscious Mind*. Oxford: Oxford University Press.

Chalmers, D. (2010) *The Character of Consciousness*. Oxford: Oxford University Press.

Chemero, A. (2009) *Radical Embodied Cognitive Science*. Cambridge, MA: MIT Press.

Chomsky, N. (1957) *Syntactic Structures*. The Hague: Mouton.

Chomsky, N. (1959) 'Review of Skinner's *Verbal Behavior*', *Language* 35: 26–58.

Chomsky, N. (1965) *Aspects of the Theory of Syntax*. Cambridge, MA: MIT Press.

Chomsky, N. (1970) 'Remarks on Nominalization', in R. Jacobs and P. Rosenbaum (eds), *Readings in English Transformational Grammar*. Waltham, MA: Ginn.

Chomsky, N. (1972) *Language and Mind*. New York: Harcourt, Brace, Jovanovich.

Chomsky, N. (1980) *Rules and Representations*. New York: Columbia University Press.

Chomsky, N. (1981) *Lectures on Government and Binding*. Hawthorne, NY: Walter de Gruyter.

Chomsky, N. (1986) *Knowledge of Language*. Westport, CT: Praeger.

Chomsky, N. (1995) *The Minimalist Program*. Cambridge, MA: MIT Press.

Chomsky, N. (2000) *New Horizons in the Study of Language and Mind*. Cambridge: Cambridge University Press.

Chomsky, N. (2002) *On Nature and Language*. Cambridge: Cambridge University Press.

Churchland, P. M. (1979) *Scientific Realism and the Plasticity of Mind*. Cambridge: Cambridge University Press.

Churchland, P. M. (1989) *A Neurocomputational Perspective*. Cambridge, MA: MIT Press.

Churchland, P. M. (2005) 'Functionalism at Forty: A Critical Retrospective', *Journal of Philosophy* 102(1): 33–50; repr. in *Neurophilosophy at Work*. New York: Cambridge University Press, 2007.

Churchland, P. M. (2007) *Neurophilosophy at Work*. New York: Cambridge University Press.

Churchland, P. M. (2012) *Plato's Camera: How the Physical Brain Captures a Landscape of Abstract Universals*. Cambridge, MA: MIT Press.

Churchland, P. M. (2013) *Matter and Consciousness*. 3rd edn, Cambridge, MA: MIT Press.

Churchland, P. S. (1986) *Neurophilosophy*. Cambridge, MA: MIT Press.

Churchland, P. S. (2013) *Touching a Nerve: The Self as Brain*. New York: W. W. Norton.

Clahsen, H., and Almazan, M. (1998) 'Syntax and Morphology in Williams Syndrome', *Cognition* 68(3): 167–98.

Clark, A. (2006) 'Vision as Dance? Three Challenges for Sensorimotor Contingency Theory', *Psyche* 12(1); www.theassc.org/files/assc/2629.pdf.

Clark, A. (2008) *Supersizing the Mind: Embodiment, Action, and Cognitive Extension*. New York: Oxford University Press.

Clark, A. (2013) 'Whatever Next? Predictive Brains, Situated Agents, and the Future of Cognitive Science', *Behavioral and Brain Sciences* 36(3): 181–204.

Clark, A. (2014) *Mindware: An Introduction to the Philosophy of Cognitive Science*. 2nd edn, Oxford: Oxford University Press.

Clark, A., and Chalmers, D. (1998) 'The Extended Mind', *Analysis* 58: 10–23.

Clark, A., and Toribio, J. (1994) 'Doing without Representing?' *Synthese* 101: 401–31.

Clark, E. (2009) *First Language Acquisition*. 2nd edn, Cambridge: Cambridge University Press.

Collins, J. (2004) 'Faculty Disputes', *Mind and Language* 17: 300–33.

Collins, J. (2008) *Chomsky: A Guide for the Perplexed*. London: Continuum.

Cottrell, G. W. (1990) 'Extracting Features from Faces using Compression Networks', in D. S. Touretzky, G. E. Hinton, T. J. Sejnowski and J. L. Elman (eds), Connectionist Models: Proceedings of the 1990 Summer School. San Mateo, CA: Morgan Kaufmann.

Cowey, A. (2004) 'Fact, Artefact and Myth about Blindsight', *Quarterly Journal of Experimental Psychology* 57A: 577–609.

Cowie, F. (1999) *What's Within: Nativism Reconsidered*. New York: Oxford University Press.

Cowie, F. (2008) 'Innateness and Language', *Stanford Encyclopedia of Philosophy*, http://plato.stanford.edu/archives/sum2010/entries/innateness-language/.

Crain, S., and Pietroski, P. (2001) 'Nature, Nurture, and Universal Grammar', *Linguistics and Philosophy* 24: 139–86.

Crane, T. (1991) 'All the Difference in the World', *Philosophical Quarterly* 41: 1–25.

Crane, T. (2003) *The Mechanical Mind: A Philosophical Introduction to Minds, Machines and Mental Representation.* 2nd edn, London: Penguin.

Crick, F. H. (1994) *The Astonishing Hypothesis: The Scientific Search for the Soul.* New York: Scribners.

Croft, W. (2001) *Radical Construction Grammar: Syntactic Theory in Typological Perspective.* Oxford: Oxford University Press.

Cummins, R. (1983) *The Nature of Psychological Explanation.* Cambridge, MA: MIT Press.

Davidson, D. (1975) 'Thought and Talk', in S. Guttenplan (ed.), *Mind and Language.* Oxford: Oxford University Press.

Davidson, D. (1980) *Essays on Actions and Events.* Oxford: Oxford University Press.

Davies, M. (2010) 'Double Dissociation: Understanding its Role in Cognitive Neuropsychology', *Mind and Language* 25: 500–40.

Dennett, D. C. (1969) *Content and Consciousness.* London: Routledge & Kegan Paul.

Dennett, D. C. (1978a) 'Artificial Intelligence as Philosophy and as Psychology', in *Brainstorms.* Cambridge, MA: MIT Press.

Dennett, D. C. (1978b) 'Skinner Skinned', in *Brainstorms.* Cambridge, MA: MIT Press.

Dennett, D. C. (1978c) *Brainstorms: Philosophical Essays on Mind and Psychology.* Cambridge, MA: MIT Press.

Dennett, D. C. (1978d) 'Response to Premack and Woodruff: Does the Chimpanzee Have a Theory of Mind?', *Behavioral and Brain Science* 4: 568–70.

Dennett, D. C. (1983) 'Styles of Mental Representation', *Proceedings of the Aristotelian Society* 83: 213–26; repr. in *The Intentional Stance.* Cambridge, MA: MIT Press, 1987.

Dennett, D. C. (1987) *The Intentional Stance.* Cambridge, MA: MIT Press.

Dennett, D. C. (1991a) 'Real Patterns', *Journal of Philosophy* 88: 27–51.

Dennett, D. C. (1991b) *Consciousness Explained.* London: Penguin.

Dennett, D. C. (2013) *Intuition Pumps and Other Tools for Thinking.* London: Penguin.

Descartes, R. (1985) *The Philosophical Writings of Descartes*, trans. J. Cottingham, R. Stoothoff and P. Murdoch. Cambridge: Cambridge University Press.

Deutscher, G. (2005) *The Unfolding of Language.* London: Arrow.

Devitt, M. (2006) *Ignorance of Language.* Oxford: Oxford University Press.

Dretske, F. (1981) *Knowledge and the Flow of Information.* Cambridge, MA: MIT Press.

Dretske, F. (1988) *Explaining Behavior.* Cambridge, MA: MIT Press.

Dretske, F. (2004) 'Change Blindness', *Philosophical Studies* 120: 1–18.

Duhem, P. (1954) *The Aim and Structure of Physical Theory*, trans. P. P. Weiner. Princeton, NJ: Princeton University Press.

Dummett, M. (1993) *The Seas of Language.* Oxford: Oxford University Press.

Dupuy, J.-P. (2000) *The Mechanization of the Mind: On the Origins of Cognitive Science.* Princeton, NJ: Princeton University Press.

Elbourne, D. (2011) *Meaning: A Slim Guide to Semantics.* Oxford: Oxford University Press.

Elman, J. L. (1991) 'Distributed Representations, Simple Recurrent Networks, and Grammatical Structure', *Machine Learning* 7: 195–224.

Epley, N. (2014) *Mindwise: How We Understand what Others Think, Believe, Feel and Want.* London: Allen Lane.

Evans, G. (1985) *Collected Papers.* Oxford: Oxford University Press.

Evans, J. (1984) 'Heuristic and Analytic Processes in Reasoning', *British Journal of Psychology* 75: 451–68.

Evans, J. (1989) *Bias in Human Reasoning: Causes and Consequences.* Hove: Lawrence Erlbaum Associates.

Evans, N., and Levinson, S. C. (2009) 'The Myth of Language Universals: Language Diversity and its Importance for Cognitive Science', *Behavioural and Brain Sciences* 32: 429–92.

Evans, V. (2015) *The Language Myth: Why Language is not an Instinct*. Cambridge: Cambridge University Press.

Everett, D. (2005) 'Cultural Constraints on Grammar and Cognition in Pirahã', *Current Anthropology* 46(4): 621–46.

Everett, D. (2012) *Language: The Cultural Tool*. London: Profile Books.

Farkas, K. (2008) *The Subject's Point of View*. Oxford: Oxford University Press.

Feigl, H. (1958) 'The "Mental" and the "Physical"', in H. Feigl, M. Scriven and G. Maxwell (eds), *Concepts, Theories and the Mind–Body Problem*. Minneapolis: University of Minnesota Press.

Feldman, J. A., and Ballard, D. H. (1982) 'Connectionist Models and their Properties', *Cognitive Science* 6: 205–54.

Fitch, W. T., Hauser, M., and Chomsky, N. (2005) 'The Evolution of the Language Faculty: Clarifications and Implications', *Cognition* 97: 179–210.

Fodor, J. A. (1968a) 'The Appeal to Tacit Knowledge in Psychological Explanation', *Journal of Philosophy* 65: 627–40; repr. in *Representations*. Cambridge, MA: MIT Press, 1981.

Fodor, J. A. (1968b) *Psychological Explanation*. New York: Random House.

Fodor, J. A. (1974) 'Special Sciences', *Synthese* 28: 97–115; repr. in *Representations*. Cambridge, MA: MIT Press, 1981.

Fodor, J. A. (1975) *The Language of Thought*. New York: Thomas Y. Crowell.

Fodor, J. A. (1978) 'Propositional Attitudes', *The Monist* 61: 501–23; repr. in *Representations*. Cambridge, MA: MIT Press, 1981.

Fodor, J. A. (1981a) 'The Present Status of the Innateness Controversy', in *Representations*. Cambridge, MA: MIT Press, 1981.

Fodor, J. A. (1981b) *Representations*. Cambridge, MA: MIT Press.

Fodor, J. A. (1983) *The Modularity of Mind*. Cambridge, MA: MIT Press.

Fodor, J. A. (1987) *Psychosemantics*. Cambridge, MA: MIT Press.

Fodor, J. A. (1990) *A Theory of Content and Other Essays*. Cambridge, MA: MIT Press.

Fodor, J. A. (1998) *Concepts*. Oxford: Oxford University Press.

Fodor, J. A. (2000) *The Mind Doesn't Work That Way*. Cambridge, MA: MIT Press.

Fodor, J. A. (2001) 'Doing without What's Within: Fiona Cowie's Critique of Nativism', *Mind* 110: 99–148.

Fodor, J. A., and Lepore, E. (1996) 'The Red Herring and the Pet Fish: Why Concepts Still Can't Be Prototypes', *Cognition* 58: 253–70.

Fodor, J. A., and McLaughlin, B. (1990) 'Connectionism and the Problem of Systematicity: Why Smolensky's Solution Doesn't Work', *Cognition* 35: 183–204.

Fodor, J. A., and Pylyshyn, Z. W. (1988) 'Connectionism and Cognitive Architecture: A Critical Analysis', *Cognition* 28: 3–71.

Frege, G. ([1892] 1980) 'On Concept and Object', in P. Geach and M. Black (eds), *Translations from the Philosophical Writings of Gottlob Frege*. Rev. edn, Oxford: Blackwell.

Frege, G. ([1914] 1979) 'Logic in Mathematics', in H. Hermes (ed.), *Posthumous Writings*. Chicago: Chicago University Press.

Frith, C. (2007) *Making up the Mind: How the Brain Makes our Mental World*. Oxford: Blackwell.

Gallacher, S. (2005) *How the Body Shapes the Mind*. Oxford: Oxford University Press.

Gallistel, C. R. (1999) 'The Replacement of General-Purpose Learning Models with Adaptively Specialized Learning Modules', in M. S. Gazzaniga (ed.), *The Cognitive Neurosciences*. 2nd edn, Cambridge, MA: MIT Press.

Gallistel, C. R., and King, A. P. (2009) *Memory and the Computational Brain: Why Cognitive Science Will Transform Neuroscience*. Chichester: Wiley-Blackwell.

Gelman, S. (2003) *The Essential Child*. New York: Oxford University Press.

Gibson, J. J. (1979) *An Ecological Approach to Visual Perception*. Boston: Houghton Mifflin.

Gigerenzer, G. (2008) *Rationality for Mortals: How People Cope with Uncertainty*. New York: Oxford University Press.

Gigerenzer, G. (2014) *Risk Savvy: How to Make Good Decisions*. London: Penguin.

Gigerenzer, G., Todd, P. M., and the ABC Research Group (1999) *Simple Heuristics That Make Us Smart*. New York: Oxford University Press.

Gilbert, M. (1989) *On Social Facts*. New York: Routledge.

Goldberg, A. E. (1995) *Constructions: A Construction Grammar Approach to Argument Structure*. Chicago: University of Chicago Press.

Goldberg, A. E. (2006) *Constructions at Work: The Nature of Generalization in Language*. Oxford: Oxford University Press.

Goldman, A. I. (1989) 'Interpretation Psychologized', *Mind and Language* 4: 161–85.

Goldman, A. I. (2006) *Simulating Minds: The Philosophy, Psychology, and Neuroscience of Mindreading*. Oxford: Oxford University Press.

Goodale, M., and Milner, D. (1992) 'Separate Visual Pathways for Perception and Action', *Trends in Neurosciences* 15(1): 20–5.

Goodale, M., and Milner, D. (2004) *Sights Unseen*. Oxford: Oxford University Press.

Gopnik, A. (2003) 'The Theory Theory as an Alternative to the Innateness Hypothesis', in L. Antony and N. Hornstein (eds), *Chomsky and his Critics*. Oxford: Blackwell.

Gopnik, A., and Meltzoff, A. N. (1997) *Words, Thoughts and Theories*. Cambridge, MA: MIT Press.

Gopnik, A., Meltzoff, A. N., and Kuhl, P. K. (1999) *The Scientist in the Crib: Minds, Brains, and How Children Learn*. New York: William Morrow.

Gopnik, M. (1994) 'Impairments of Tense in a Familial Language Disorder', *Journal of Neurolinguistics* 8: 109–33.

Gordon, R. (1986) 'Folk Psychology as Simulation', *Mind and Language* 1: 158–71.

Gregory, R. (1966) *Eye and Brain: The Psychology of Seeing*. London: Weidenfeld & Nicolson.

Gregory, R. (1997) *Eye and Brain: The Psychology of Seeing*. 5th edn, Oxford: Oxford University Press.

Grice, H. P. (1957) 'Meaning', *Philosophical Review* 66: 377–88.

Grice, H. P. (1975) 'Logic and Conversation', in D. Davidson and G. Harman (eds), *The Logic of Grammar*. Encino, CA: Dickenson.

Griggs, R. A., and Cox, J. R. (1982) 'The Elusive Thematic Materials Effect in the Wason Selection Test', *British Journal of Psychology* 73: 407–20.

Hampton, J. A. (1979) 'Polymorphous Concepts in Semantic Memory', *Journal of Verbal Learning and Verbal Behavior* 18: 441–61.

Hansen, T., Olkkonen, M., Walter, S., and Gegenfurtner, K. R. (2006) 'Memory Modulates Color Appearance', *Nature Neuroscience* 9(11): 1367–8.

Harman, G. (1987) 'Conceptual Role Semantics', *Notre Dame Journal of Formal Logic* 23: 242–56.

Haugeland, J. (1981) *Artificial Intelligence: The Very Idea*. Cambridge, MA: MIT Press.

Hauser, M., Chomsky, N., and Fitch, W. T. (2002) 'The Language Faculty: What is it, Who has it, and How did it Evolve?' *Science* 298: 1569–79.

Heal, J. (1986) 'Replication and Functionalism', in J. Butterfield (ed.), *Language, Mind, and Logic*. Cambridge: Cambridge University Press.

Hebb, D. (1949) *The Organization of Behavior: A Neuropsychological Theory*. New York: Wiley.

Hinton, G. E., McClelland, J. L., and Rumelhart, D. E. (1986) 'Distributed Representations', in Rumelhart et al. (eds), *Parallel Distributed Processing*, Vol. 1: *Foundations*. Cambridge, MA: MIT Press.

Hohwy, J. (2013) *The Predictive Mind*. Oxford: Oxford University Press.

Hubel, D. H., and Wiesel, T. N. (1959) 'Receptive Fields of Single Neurons in the Cat's Striate Cortex', *Journal of Physiology* 148: 574–91.

Hubel, D. H., and Wiesel, T. N. (1962) 'Receptive Fields, Binocular Interaction and Functional Architecture in the Cat's Visual Cortex', *Journal of Physiology* 160: 106–54.

Hume, D. ([1738] 1978) *A Treatise of Human Nature*, ed. L. A. Selby-Bigge. 2nd edn, rev. P. H. Nidditch. Oxford: Clarendon Press.

Hurley, S. (1998) *Consciousness in Action*. Cambridge, MA: Harvard University Press.

Hutto, D., and Myin, E. (2013) *Radicalizing Enactivism*. Cambridge, MA: MIT Press.

Jackendoff, R. (1977) *X-Bar Syntax: A Theory of Phrase Structure*. Cambridge, MA: MIT Press.

Jackendoff, R. (2002) *Foundations of Language*. Oxford: Oxford University Press.

Jackendoff, R. (2012) *A User's Guide to Thought and Meaning*. Oxford: Oxford University Press.

Jackson, F. (1986) 'What Mary Didn't Know', *Journal of Philosophy* 83: 209–25.

Jacob, F. (1977) 'Evolution and Tinkering', *Science* 196: 1161–6.

Jacob, P. (2006) 'Why Visual Experience is Likely to Resist Being Enacted', *Psyche* 12(1); www.theassc.org/files/assc/2630.pdf.

James, W. ([1890] 1995) *The Principles of Psychology*. New York: Dover.

Jenkins, L. (2000) *Biolinguistics: Exploring the Biology of Language*. Cambridge: Cambridge University Press.

Kahneman, D. (2012) *Thinking, Fast and Slow*. London: Penguin.

Kandel, E. R., Schwartz, J. H., Jessell, T. M., Siegelbaum, S. A., and Hudspeth, A. J. (2012) *Principles of Neural Science*. 5th edn, New York: McGraw-Hill.

Kant, I. ([1781] 1998) *Critique of Pure Reason*, trans. and ed. P. Guyer and A. Wood. Cambridge: Cambridge University Press.

Karmiloff-Smith, A. (1996) *Beyond Modularity: A Developmental Perspective on Cognitive Science*. Cambridge, MA: MIT Press.

Karmiloff-Smith, A., Klima, E., Grant, J., and Baron-Cohen, S. (1995) 'Is there a Social Module? Language, Face Processing and Theory of Mind in Individuals with Williams Syndrome', *Journal of Cognitive Neuroscience* 7: 196–208.

Keijzer, F. (1999) *Representation and Behavior*. Cambridge, MA: MIT Press.

Keil, F. (1989) *Concepts, Kinds, and Cognitive Development*. Cambridge, MA: MIT Press.

Kenny, A. (1984) *The Legacy of Wittgenstein*. Oxford: Blackwell.

Koch, C. (2004) *The Quest for Consciousness: A Neurobiological Approach*. Denver: Roberts.

Kripke, S. (1980) *Naming and Necessity*. Oxford: Blackwell.

Lakoff, G. (1987) *Women, Fire and Dangerous Things: What Categories Reveal about the Mind.* Chicago: University of Chicago Press.

Langacker, R. W. (1987) *Foundations of Cognitive Grammar*, Vol. 1: *Theoretical Prerequisites.* Stanford, CA: Stanford University Press.

Langacker, R. W. (1991) *Foundations of Cognitive Grammar*, Vol. 2: *Descriptive Application.* Stanford, CA: Stanford University Press.

Larson, R. K. (2010) *Grammar as Science.* Cambridge, MA: MIT Press.

Laurence, S., and Margolis, E. (2001) 'The Poverty of the Stimulus Argument', *British Journal for the Philosophy of Science* 52: 217-76.

Levin, D. T., Drivdahl, S. B., Momen, N., and Beck, M. R. (2002) 'False Predictions about the Detectability of Visual Changes: The Role of Beliefs about Attention, Memory, and the Continuity of Attended Objects in Causing Change Blindness Blindness', *Consciousness and Cognition* 11: 507-27.

Levine, J. (2001) *Purple Haze: The Puzzle of Consciousness.* New York: Oxford University Press.

Lewis, D. (1969) *Convention: A Philosophical Study.* Cambridge, MA: Harvard University Press.

Lewis, D. (1975) 'Language and Languages', in K. Gunderson (ed.), *Language, Mind and Knowledge.* Minneapolis: University of Minnesota Press; repr. in *Philosophical Papers*, Vol. 1. New York: Oxford University Press, 1983.

Locke, J. ([1689] 1975) *An Essay Concerning Human Understanding*, ed. P. H. Nidditch. Oxford: Clarendon Press.

McClelland, J. L., Rumelhart, D. E., and the PDP Research Group (eds) (1986) *Parallel Distributed Processing: Explorations in the Microstructures of Cognition*, Vol. 2: *Psychological and Biological Models.* Cambridge, MA: MIT Press.

McDowell, J. (1994) *Mind and World.* Cambridge, MA: Harvard University Press.

McGilvray, J. (2005) *The Cambridge Companion to Chomsky.* Cambridge: Cambridge University Press.

McGilvray, J. (2013) *Chomsky: Language, Mind and Politics.* 2nd edn, Cambridge: Polity.

McGinn, C. (1989a) *Mental Content.* Oxford: Blackwell.

McGinn, C. (1989b) 'Can We Solve the Mind–Body Problem?' *Mind* 98: 349-66.

Machery, E. (2009) *Doing without Concepts.* New York: Oxford University Press.

Marcus, G. (2001) *The Algebraic Mind: Integrating Connectionism and Cognitive Science.* Cambridge, MA: MIT Press.

Margolis, E., and Laurence, S. (2007) 'The Ontology of Concepts - Abstract Objects or Mental Representations?' *Nous* 41(4): 561-93.

Marr, D. (1982) *Vision.* San Francisco: W. H. Freeman.

Marslen-Wilson, W. D., and Tyler, L. K. (1987) 'Against Modularity', in J. L. Garfield (ed.), *Modularity in Knowledge Representation and Natural Language Understanding.* Cambridge, MA: MIT Press.

Medin, D., and Ortony, A. (1989) 'Psychological Essentialism', in S. Vosniadou (ed.), *Similarity and Analogical Reasoning.* New York: Cambridge University Press.

Medin, D. L., and Shaffer, M. M. (1978) 'Context Theory of Classification Learning', *Psychological Review* 85: 207-38.

Menary, R. (2007) *Cognitive Integration: Mind and Cognition Unbounded.* Basingstoke: Palgrave Macmillan.

Menary, R. (ed.) (2010) *The Extended Mind.* Cambridge, MA: MIT Press.

Miller, G. (1956) 'The Magical Number Seven, Plus or Minus Two: Some Limits on Our Capacity for Processing Information', *Psychological Review* 63(2): 81–97.

Millikan, R. G. (1984) *Language, Thought and Other Biological Categories*. Cambridge, MA: MIT Press.

Millikan, R. G. (1993) *White Queen Psychology and Other Essays for Alice*. Cambridge, MA: MIT Press.

Millikan, R. G. (1998) 'Language Conventions Made Simple', *Journal of Philosophy* 95: 161–80.

Millikan, R. G. (2003) 'In Defense of Public Language', in L. Antony and N. Hornstein (eds), *Chomsky and his Critics*. Oxford: Blackwell.

Minsky, M., and Papert, S. (1969) *Perceptrons: An Introduction to Computational Geometry*. Cambridge, MA: MIT Press.

Murphy, D. (2002) *The Big Book of Concepts*. Cambridge, MA: MIT Press.

Nagel, N. (1974) 'What Is it Like to Be a Bat?' *Philosophical Review* 83: 435–50.

Nanay, B. (2013) *Between Perception and Action*. Oxford: Oxford University Press.

Newell, A., and Simon, H. (1961) 'Simulation of Human Thought', in D. Wayne (ed.), *Current Trends in Psychological Theory*. Pittsburgh: University of Pittsburgh Press.

Newell, A., and Simon, H. (1976) 'Computer Science as Empirical Enquiry: Symbols and Search', *Communications of the Association for Computing Machinery* 19: 113–26.

Newport, E. L., Gleitman, H., and Gleitman, L. A. (1977) 'Mother, I'd Rather Do it Myself: Some Effects and Non-Effects of Maternal Speech Style', in C. E. Snow and C. A. Ferguson (eds), *Talking to Children: Language Input and Acquisition*. Cambridge: Cambridge University Press.

Noë, A. (2004) *Action in Perception*. Cambridge, MA: MIT Press.

O'Regan, K., and Noë, A. (2001) 'A Sensorimotor Account of Vision and Visual Consciousness', *Behavioral and Brain Sciences* 24(5): 939–1031.

Papineau, D. (1993) *Philosophical Naturalism*. Oxford: Blackwell.

Papineau, D. (2012) *Philosophical Devices: Proofs, Probabilities, Possibilities, and Sets*. Oxford: Oxford University Press.

Perner, J. (1991) *Understanding the Representational Mind*. Cambridge, MA: MIT Press.

Pinker, S. (1989) *Learnability and Cognition: The Acquisition of Argument Structure*. Cambridge, MA: MIT Press.

Pinker, S. (1994) *The Language Instinct: The New Science of Language and Mind*. London: Penguin.

Pinker, S. (1997) *How The Mind Works*. New York: W. W. Norton.

Pinker, S. (2007) *The Stuff of Thought*. London: Allen Lane.

Pinker, S., and Bloom, P. (1990) 'Natural Language and Natural Selection', *Behavioural and Brain Sciences* 13: 707–84.

Place, U. T. (1956) 'Is Consciousness a Brain Process?' *British Journal of Psychology* 47: 44–50.

Popper, K. R. (1959) *The Logic of Scientific Discovery*. London: Hutchinson.

Prinz, J. (2002) *Furnishing the Mind: Concepts and their Perceptual Basis*. Cambridge, MA: MIT Press.

Prinz, J. (2005) 'The Return of Concept Empiricism', in H. Cohen and C. Lefebvre (eds), *Categorization and Cognitive Science*. Amsterdam: Elsevier.

Prinz, J. (2006a) 'Is the Mind Really Modular?' in R. A. Stainton (ed.), *Contemporary Debates in Cognitive Science*. Oxford: Blackwell.

Prinz, J. (2006b) 'Putting the Brakes on Enactive Perception', *Psyche* 12(1); www.theassc. org/files/assc/2627.pdf.

Prinz, J. (2012) *Beyond Human Nature: How Culture and Experience Shape our Lives.* London: Penguin.

Pullum, G. K., and Scholz, B. C. (2002) 'Empirical Assessment of Stimulus Poverty Arguments', *Linguistic Review* 19: 9–50.

Putnam, H. (1967) 'Psychological Predicates', in W. H. Capitan and D. D. Merrill (eds), *Art, Mind, and Religion.* Pittsburgh: University of Pittsburgh Press.

Putnam, H. (1975) 'The Meaning of "Meaning"', in *Mind, Language and Reality: Philosophical Papers*, Vol. 2. Cambridge: Cambridge University Press.

Pylyshyn, Z. W. (1984) *Computation and Cognition.* Cambridge, MA: MIT Press.

Quine, W. V. O. (1951) 'The Two Dogmas of Empiricism', *Philosophical Review* 60: 20–43; repr. in Quine, *From a Logical Point of View.* Cambridge, MA: Harvard University Press, 1953.

Radford, A. (2009) *An Introduction to English Sentence Structure.* Cambridge: Cambridge University Press.

Ramachandran, V. S. (2011) *The Tell-Tale Brain: Unlocking the Mystery of Human Nature.* London: William Heinemann.

Rey, G. (1994) 'Concepts', in S. D. Guttenplan (ed.), *A Companion to the Philosophy of Mind.* Oxford: Blackwell.

Rips, L. J., Shoben, E. J., and Smith, E. E. (1973) 'Semantic Distance and the Verification of Semantic Relations', *Journal of Verbal Learning and Verbal Behavior* 12: 1–20.

Rizzolatti, G., Fadiga, L., Fogassi, L., and Gallese, V. (1996) 'Premotor Cortex and the Recognition of Motor Actions', *Cognitive Brain Research* 3: 131–41.

Rosch, E. (1978) 'Principles of Categorization', in E. Rosch and B. B. Lloyd (eds), *Cognition and Categorization.* Hillsdale, NJ: Lawrence Erlbaum.

Rosenblatt, F. (1962) *Principles of Neurodynamics.* Washington, DC: Spartan Books.

Rowland, C. (2014) *Understanding Child Language Acquisition.* Abingdon: Routledge.

Rowlands, M. (2003) *Externalism: Putting Mind and World Back Together Again.* Chesham: Acumen.

Rumelhart, D. E., and McClelland, J. L. (1986) 'On Learning the Past Tense of English Verbs', in Rumelhart et al. (eds), *Parallel Distributed Processing*, Vol. 1: *Foundations.* Cambridge, MA: MIT Press.

Rumelhart, D. E., McClelland, J. L., and the PDP Research Group (eds) (1986) *Parallel Distributed Processing: Explorations in the Microstructures of Cognition*, Vol. 1: *Foundations.* Cambridge, MA: MIT Press.

Rupert, R. (2009) *Cognitive Systems and the Extended Mind.* Oxford: Oxford University Press.

Russell, J. (2004) *What is Language Development? Rationalist, Empiricist and Pragmatist Approaches to the Acquisition of Syntax.* Oxford: Oxford University Press.

Ryle, G. (1949) *The Concept of Mind.* Harmondsworth: Penguin.

Saffran, J. R., Aslin, R. N., and Newport, E. L. (1996) 'Statistical Learning by 8-Month-Old Infants', *Science* 274: 1926–8.

Sampson, G. (2005) *The Language Instinct Debate.* London: Continuum.

Samuels, R. (2006) 'Is the Human Mind Massively Modular?' in R. A. Stainton (ed.), *Contemporary Debates in Cognitive Science.* Oxford: Blackwell.

Schiffer, S. R. (1987) *Remnants of Meaning.* Cambridge, MA: MIT Press.

Schneider, S. (2011) *The Language of Thought: A New Philosophical Direction.* Cambridge, MA: MIT Press.

Scholl, B. J., and Leslie, A. M. (1999) 'Modularity, Development and Theory of Mind', *Mind and Language* 14: 131–53.

Scholz, B., and Pullum, G. (2006) 'Irrational Nativist Exuberance', in R. A. Stainton (ed.), *Contemporary Debates in Cognitive Science*. Oxford: Blackwell.

Searle, J. R. (1980) 'Minds, Brains and Programs', *Behavioral and Brain Sciences* 3: 417–24.

Searle, J. R. (1983) *Intentionality: An Essay in the Philosophy of Mind*. Cambridge: Cambridge University Press.

Searle, J. R. (1992) *The Rediscovery of the Mind*. Cambridge, MA: MIT Press.

Segal, G. (1996) 'The Modularity of Theory of Mind', in P. Carruthers and P. K. Smith (eds), *Theories of Theories of Mind*. Cambridge: Cambridge University Press.

Shapiro, L. (2011) *Embodied Cognition*. New York: Routledge.

Simon, H. (1962) 'The Architecture of Complexity', *Proceedings of the American Philosophical Society* 106(6): 467–82.

Simons, D., and Chabris, C. (1999) 'Gorillas in our Midst: Sustained Inattentional Blindness for Dynamic Events', *Perception* 28: 1059–74.

Skinner, B. F. (1953) *Science and Human Behavior*. New York: Macmillan.

Skinner, B. F. (1957) *Verbal Behaviour*. London: Methuen.

Sloman, S. (1996) 'The Empirical Case for Two Systems of Reasoning', *Psychological Bulletin* 119: 3–22.

Smart, J. J. C. (1959) 'Sensations and Brain Processes', *Philosophical Review* 68: 141–56.

Smith, E. E., and Kosslyn, S. M. (2007) *Cognitive Psychology: Mind and Brain*. Upper Saddle River, NJ: Pearson.

Smith, N. V., and Tsimpli, I. M. (1995) *The Mind of a Savant: Language Learning and Modularity*. Oxford: Blackwell.

Smith, N. V., Tsimpli, I. M., Morgan, G., and Woll, B. (2010) *Signs of a Savant: Language against the Odds*. Cambridge: Cambridge University Press.

Smolensky, P. (1991) 'Connectionism, Constituency and the Language of Thought', in B. Loewer and G. Rey (eds), *Meaning in Mind: Fodor and his Critics*. Oxford: Blackwell.

Snow, C. E. (1977) 'The Development of Conversation Between Mothers and Babies', *Journal of Child Language* 4: 1–22.

Spelke, E. (1994) 'Initial Knowledge: Six Suggestions', *Cognition* 50: 435–45.

Spelke, E. (2003) 'What Makes Us Smart? Core Knowledge and Natural Language', in D. Gentner and S. Goldin-Meadow (eds), *Language in Mind*. Cambridge, MA: MIT Press.

Sperber, D. (1996) *Explaining Culture: A Naturalistic Approach*. Oxford: Blackwell.

Squire, L. R., Bloom, F. E., Spitzer, N. C., du Lac, S., Ghosh, A., and Berg, D. (2013) *Fundamental Neuroscience*. Burlington, MA: Academic Press.

Stanovich, K. (1999) *Who is Rational? Studies of Individual Differences in Reasoning*. Mahwah, NJ: Lawrence Erlbaum Associates.

Sterelny, K. (2003) *Thought in a Hostile World: The Evolution of Human Cognition*. Oxford: Blackwell.

Stone, T., and Davies, M. (2012) 'Theoretical Issues in Cognitive Psychology', in N. Braisby and A. Gellatly (eds), *Cognitive Psychology*. 2nd edn, Oxford: Oxford University Press.

Thelen, E., and Smith, L. B. (1994) *A Dynamic Systems Approach to the Development of Cognition and Action*. Cambridge, MA: MIT Press.

Thompson, E. (2007) *Mind and Life*. Cambridge, MA: Harvard University Press.

Tomasello, M. (2003) *Constructing a Language: A Usage-Based Theory of Language Acquisition*. Cambridge, MA: MIT Press.

Tomasello, M. (2008) *Origins of Human Communication*. Cambridge, MA: MIT Press.

Tooby, J., and Cosmides, L. (1992) 'The Psychological Foundations of Culture', in J. Barkow, L. Cosmides and J. Tooby (eds), *The Adapted Mind: Evolutionary Psychology and the Generation of Culture*. New York: Oxford University Press.

Tronick, E., Als, H., Adamson, L., Wise, S., and Brazelton, T. B. (1978) 'The Infant's Response to Entrapment Between Contradictory Messages in Face-to-Face Interaction', *Journal of the American Academy of Child and Adolescent Psychiatry* 17: 1–13.

Ungerleider, L. G., and Mishkin, M. (1982) 'Two Cortical Visual Systems', in D. J. Ingle, M. A. Goodale and R. J. W. Mansfield (eds), *Analysis of Visual Behaviour*. Cambridge, MA: MIT Press.

van Gelder, T. (1995) 'What Might Cognition Be, if Not Computation?' *Journal of Philosophy* 92(7): 345–81.

Varela, F., Thompson, E., and Rosch, E. (1991) *The Embodied Mind: Cognitive Science and Human Experience*. Cambridge, MA: MIT Press.

Vargha-Khadem, F., Watkins, K. E., Price, C. J., Ashburner, J., Alcock, K. J., Connelly, A., Frackowiak, R. S., Friston, K. J., Pembrey, M. E., Mishkin, M., Gadian, D. G., and Passingham, R. E. (1998) 'Neural Basis of an Inherited Speech and Language Disorder', *Proceedings of the National Academy of Sciences* 95(21): 12695–700.

Wason, P., and Johnson-Laird, P. (1972) *Psychology of Reasoning: Structure and Content*. Cambridge, MA: Harvard University Press.

Watson, J. B. (1913) 'Psychology as the Behaviorist Views it', *Psychological Review* 20: 158–77.

Weiskrantz, L. (1986) *Blindsight: A Case Study and Implications*. Oxford: Oxford University Press.

Wheeler, M. (2005) *Reconstructing the Cognitive World*. Cambridge, MA: MIT Press.

Wilson, R. A. (2004) *Boundaries of the Mind: The Individual in the Fragile Sciences*. Cambridge: Cambridge University Press.

Wilson, R. A., and Clark, A. (2009) 'How to Situate Cognition: Letting Nature Take its Course', in M. Aydede and P. Robbins (eds), *The Cambridge Handbook of Situated Cognition*. Cambridge: Cambridge University Press.

Wimmer, H., and Perner, J. (1983) 'Beliefs about Beliefs: Representation and Constraining Function of Wrong Beliefs in Young Children's Understanding of Deception', *Cognition* 13(1): 103–28.

Wittgenstein, L. (1953) *Philosophical Investigations*, trans. G. E. M. Anscombe. 3rd edn, Oxford: Blackwell.

Wynn, K. (1992) 'Addition and Subtraction by Human Infants', *Nature* 358: 749–50.

Zeki, S. (1993) *A Vision of the Brain*. Oxford: Blackwell.

Zeman, A. (2008) *A Portrait of the Brain*. New Haven, CT: Yale University Press.

Index

Action in Perception (Noë) 187
The Aim and Structure of Physical Theory (Duhem) 11
Ames room 68
animals
  cognition and 61–2
  mental states of 165–6
  propositional attitudes and 36–7
Artificial Intelligence
  interdisciplinarity 23
  multiple realizability 163
  reasoning and 38
Aspects of the Theory of Syntax (Chomsky) 125
atomism
  concepts and 96–8
  informational 98, 112–13
attention
  change blindness 1801, 190, 192
  inattentional blindness 180–1, 188, 190, 192
Attention Deficit Hyperactivity Disorder (ADHD) 11–13

Baillargeon, Renée 156
Bayes's (Thomas) theorem 182–5, 198
Bayne, T. 176
beliefs
  concepts and 95–6
  modularity and 76–8
  physicalism and 20
  propositional attitudes and 2–3
  scientific reasoning and 80–1
Bellugi, Ursula 73
Bickle, John 165
blindness
  agnosia 179
  blind-sight 178
  change blindness 182, 191, 193
  enactivism and 190

  inattentional blindness 181–2, 188, 191, 193
  optic ataxia 179
Bloom, P. 86
  children and essence 122
  language acquisition 142
Bloomfield, Leonard 124
Boden, Margaret 138
body
  mind–body dualism 6–7, 19–20, 21–2, 58
  see also brain; neuroscience
Boyle, Robert
  natural laws 10
the brain 173
  biolinguistics and 125–7
  brainstem and cerebellum 173
  cerebral cortex of 170–1
  complexity of 161–2
  computers and 54
  enactivism and 187–94
  face identification 178
  Hebbian learning 173
  limbic system of 172–3
  lobes of 171–4
  as mind 162–3
  phantom sensations 173–4
  role of neurons 167–70
  scanning of 174–5, 179–80
  somatic/automatic nervous systems 167–8
  structure and function of 167–75
  visual processes of 176–8
Braine, M. D. S. 138
Broca, Pierre Paul 20
Brooks, Rodney 60–1, 191
Brown, R. 138

Campos, J. 155
Carey, Susan 89, 110
Carruthers, Peter 67, 79
causality
  networks of representation 114
  theory theory of concepts 111–12

Chabris, C.
  inattentional blindness 181–2,
    193
  multi-tasking and language 140–1
  perception and 70
Chalmers, David 186
  'The Extended Mind' (with Clark)
    62–4
change blindness *see* blindness
children
  Child Directed Speech/Motherese
    144
  infants and objects 84–9
  language acquisition 135–8,
    141–6
  Tomasello on language 149–59
Chomsky, Noam
  Aspects of the Theory of Syntax
    125
  commonsense selves 8–9
  competence *versus* performance
    73–4, 139
  critique of Skinner 16–17, 124
  E-language/I-language 125
  generative grammar 27, 75, 159
  interdisciplinarity 24
  language acquisition 124, 126–8,
    134–8
  modularity and 66
  objections to 138–46, 159–60
  problems and mysteries 7
  Syntactic Structures 124
  Tomasello and 150, 154
Churchland, Paul M.
  colour perception 52–3
  connectionism and 42, 44, 46–53,
    56–7
  Hebbian learning 173
  recurrent networks 51–2
Clark, Andy 188
  'The Extended Mind' (with
    Chalmers) 62–4
  four E approach 57
  representation and 62
classical computationalism *see*
    computationalism
cognition
  animal experiments and 165–6
  anti-representational approaches
    57–62, 64
  brain damage and 178
  defining 1–5
  degradation of 54–5
  Descartes and 57–8
  the extended mind 62–5

four Es of 57, 196
  machines and robots 58–61
  mental processes of 3–4
  multiple realizability 163–4
  rational coherence 36
  as theoretical entity 11
  thinking and 1–2
  the unconscious and 4
  various representational
    approaches 57–61
cognitive science
  commonsense and 7–9
  controversies of 195–8
  defining 1
  interdisciplinarity 23–4
  mental representations 17–19
  neuroscience and 161–7
  origins of 14–23
  philosophy of 1
  scepticism and 6–9
  scientific investigation 6
cognitive subtraction 175
commonsense
  cognitive science and 195
  mind and 1
  science and 13–14
communication
  beyond literal understanding 8–9
  common ground and 149
  higher-order mental states 148
  non-verbal 148–9
  see also language and linguistics
computationalism
  Chomsky and 27
  Churchland's objections 36–7
  classical approach of 25–35
  computer language and 28–35
  connectionism and 41–2, 53
  controversies of 195, 196
  Dennett and 26, 32–3
  dynamic thought and 197
  enactivism and 188
  evaluating approach of 35–41
  the extended mind 64
  Fodor and 34–5
  Marr and 32, 33–4
  massive modularity and 79
  neuroscience and 162
  propositional attitudes 35
  reasoning and 37–40
  representation and 25, 40–1
  Schneider and 31
  see also connectionism
computers *see* Artificial Intelligence;
    computationalism

concepts
  abstract 118–19
  acquisition of 103–4
  atomism and holism 96–8
  categories and inference 115
  classical theory 100–1
  exemplar theory 98, 104–5
  imagistic theory 99–100
  informational atomism 98
  language and 94–5
  prototype theory 98, 101–8
  proxytype theory 98, 113–22
  representation 99
  rich/thin 96
  sharing 112
  theories of 98–101
  theory theory 98
  unchangingness of 95
  units of thought 93–5
Concepts (Fodor) 100–1
connectionism
  brain speed and 54
  Churchland and 44, 46–53
  and classical computationalism
    53
  colour perception 52–3
  controversies of 195
  degradation of cognition and
    54–5
  emergence of 25
  evaluation of 53–7
  excitatory and inhibitory
    connections 43–4
  holism and 50
  learning and 56–7
  networks of 42–4, 51–2
  relation to computationalism
    41–2
  systematicity of thought 55–6
consciousness
  phenomenal 7
  see also the unconscious
conservatism, principle of 77
Constructing and Language
  (Tomasello) 146
Cosmides, L.
  biological mind 78–9
  cheater detection module
    81–3
Cottrell, G. W. 44–6, 47–9, 54
Cowie, Fiona
  impairment and language 73,
    74
  language acquisition 142–3,
    144–6

Cox, J. R. 82
Crain, S. 143
cybernetics 57

Dennett, Daniel
  computers as syntactic engine
    26
  personal and subpersonal
    32–3
Descartes, René
  cognition 57–8
  mind–body dualism 6–7, 19–20,
    21–2, 58
desires, physicalism and 20
Deutscher, G. 138
dual aspect theories 39–40
Duhem, Pierre
  *The Aim and Structure of Physical
    Theory* 11
  Duhem–Quine thesis 11–13

emotions
  limbic system of brain and
    172–3
  moods 1
  propositional attitudes and 3
enactivism
  blindness and 188, 190–1
  Noë's theory of 187–91
  objections to 191–4
  representations and 193–4
Epley, N. 86, 89
essentialism
  children and 110, 122
  defining an essence 110
  psychological 120–1
  real *versus* nominal 120
  Twin Earth thought experiment
    120–2
Evans, N. 141–2
Everett, Daniel 138–41
evolution
  acquiring knowledge and 84
  of cognitive systems 58
  exemplar theory 104–5
  concepts and 98
'The Extended Mind' (Clark and
    Chalmers) 62–4

face-recognition network
  connectionism and 44–6, 47–9,
    54
Feigl, H. 162
Fitch, W. T. 139
Fletcher, Dee 178–80

Fodor, Jerry
  atomism 97
  on causal role semantics 111–12
  the central system and modularity
    75–8
  Concepts 100–1
  defining modules 66–8
  geological laws 10
  language of thought 48, 112–13
  lexical concepts 115–16
  The Modularity of Mind 66–70
  objections to prototype theory
    105–7
  propositional attitudes 34–5, 35
  scientific confirmation 79–80
  simple and complex symbols 117
  systematicity of thought 55–6
Frege, Gottlob 93–5, 98
Freud, Sigmund 15, 22–3
Frith, Chris 186–7
functional role theories 41
functionalism
  Turing machines 22
  type-type identity theory 21–2

Galileo Galilei 9, 10
Gallistel, C. R. 37
games, Wittgenstein and 100
Gibson, J. J. 57, 191, 196
Gigerenzer, Gerd 39, 81
Goldman, A. I. 90
Goodale, M. 178–80
Gopnik, Alison 89, 144
Gopnik, Myrna 74
Gordon, R. 89
Gregory, R. 188
Grice, H. P. 148, 155
Griggs, R. A. 82

Hansen, T. 72, 181
Haugeland, John 29–30
Hauser, M. 139
Heal, J. 89
Hebb, Donald 56–7, 173
Heidegger, Martin 57, 196
Herring illusions 68
heuristics 197
  reasoning and 39–40
  scientific reasoning and 81
holism
  concepts and 96–8
  connectionist networks 50
  modularity and 78
Hubel, David H. 165, 177
Hume, David

concept acquisition 106
imagistic theory 99
mental representation 17
A Treatise of Human Nature 14
Hurley, S. 188

imagistic theory 99–100
inattentional blindness *see* blindness
informational atomism 98, 112–13
informational theories
  computationalism and 41
intentionality
  computationalism and 40

Jackendoff, Ray 64–5, 123
Jacob, François 58
James, William
  on attention 181
  mental representation 17
Johnson-Laird, P. 81–2, 83

Kahneman, Daniel
  Thinking, Fast and Slow 38–9, 40
Kant, Immanuel
  mental representation 17
  a priori perception 186
Karmiloff-Smith, Annette
  modularity 67
  on Williams syndrome 73
Keil, Frank 110
Kepler, Johannes 10
knowledge
  evolution and 84
  of language 123–5
Kosslyn, S. M. 70
Kuhl, P. K. 144

language and linguistics
  abstract 150–4
  acquisition 124–8, 134–8,
    146–59
  beyond literal understanding 8–9
  biolinguistics 125–7
  Chomsky's generative grammar
    75
  compared to mathematics 156–8
  competence *versus* performance
    73–4
  computer 25–6, 28–35
  construction grammar 150
  controversies of 195
  critique of Chomsky 141–6
  differences of languages 141–2
  environmental factors 143–4
  generative grammar 159

hierarchy of representation 129–8
humans and animals 35–6
importance in cognitive science 123
interdisciplinarity 23–4
knowledge of 123–6
language of thought (LOT) 34–5, 48, 112–13
learning 16–17
lexical concepts 115–16
linguistic creativity 129
manifesting thought 64–5
meaning and 28
mind and 24
modularity and 90
multi-tasking and 140–1
natural 26–9
oral languages 139
pattern-finding and mind-reading 147–8, 151–59
Pirahã language 138–41
primary linguistic data (pld) 135
propositional attitudes and 2–3
psychology and 124–5
relation to concepts 94–5
representations 126
social structures and 124–5, 138–41
Specific Language Impairment 74
syntax 74–5
systematicity of thought 55–6
universals 128, 136–8, 141–2, 198
laws, natural 9–10
learning
connectionism and 56–7
Hebbian 173
mathematics *versus* language 156–8
stimulus–response 15–16
see also children; psychology
Levinson, S. C. 141–2
linguistics *see* language and linguistics
Locke, John
concept acquisition 106
imagistic theory 99
primary and secondary qualities 186
real *versus* nominal essences 120
senses first 115

McClelland, J. L. 25
McGinn, Colin 7
Machery, Edouard 105–6

machines
cognition and 61
see also Artificial Intelligence; robots
Marr, David
classical computationalism 32
connectionism and representation 53
enactivism and 188
levels of cognition 163–4
neuroscience and 180
representations 193
theory of vision 33–4, 167, 197
Marslen-Wilson, W. D. 74–5
massive modularity thesis *see* modularity
mathematics learning 156–8
Medin, D. 110
Meltzoff, A. N. 89, 144
memory
limbic system of brain and 172–3
proxytype theory 113–22
representations 114
writing and 65
mental states *see* mind
Merleau-Ponty, Maurice 57, 196
metaphysics of mind
type-type identity theory 20–1
Milner, D. 178–80
mind
biolinguistics and 125–7
as brain function 162–3
Darwinian view of 78–9
extended 62–5
Freud's mechanisms of 22–3
higher order mental states 148
modularity subsystems 67–8
reading other people's 89–92, 98, 147–8, 151–59
Turing and 18–19
type-type identity 20–1
see also the brain; mind–body dualism
mind–body dualism
degradation of cognition 54–5
Descartes and 6–7, 58
type-type identity theory 20–2
mind-reading 89–92, 98
language and 147–8, 151–59
Mishkin, M. 177–8
modularity
beliefs and 76–8
the central system 75–8
cheater detection module 81–3
complex tasks and 83–4

modularity (*cont.*)
  controversies of 196
  development of thesis 66
  Fodorian modules 66–75
  holism and 78
  infants and objects 84–9
  language and 73–5
  massive 78–84, 197
  perception and 68–72
  predicting behaviour of others
    89–92
  Prinz's objections to 71, 74–5
  subsystems 67–8
  theory of mind 90
The Modularity of Mind (Fodor)
  66–70
moods, cognition and 1
Müller–Lyer illusion 68
multiple realizability 163–5
Murphy, D. 103

nature *see* laws, natural
networks *see* connectionism:
    networks
neuroscience
  animal experiments and 165–6
  brain structure and function
    167–75
  cognitive science and 161–7
  colour perception 180–1
  interdisciplinarity 23
  multiple realizability 163–5
  neuropsychology and modularity
    72–3
Newport, E. L. 144
Newton, Isaac 10
Noë, Alva
  Action in Perception 187
  enactivism theory 187–94

Origins of Human Communication
  (Tomasello) 146
Ortony, A. 110
other minds 89–92, 98

pain
  type-type identity theory 21
Parallel Distributed Processing *see*
    connectionism
pattern-finding 151–59
  language and 147–8
perception
  behaviour of objects 91
  of colour 52–3, 69, 72, 180–1
  controversies of 197–8

  definition 3
  illusions 68–9
  independence from mind 186–7
  modularity thesis 68–72
  neuroscience of 178–85
  propositional attitudes and 3, 5
  sense and 115
  shapes and 189–90
  vision and 3
phenomenology
  imagistic theory and 99
  of vision 192–3
philosophy
  of cognitive science 1
  concept acquisition 105–6
  interdisciplinarity 23
physicalism 19, 164
  type-type 20–1
Pietroski, P. 143
Pinker, Steven
  language acquisition 137, 138,
    142
  Williams syndrome and language
    73
Pirahã language 138–41
Place, U. T. 162
Popper, Karl 10–11
primary linguistic data (pld) 135–9,
    143–5, 147
Prinz, Jesse
  objections to modularity 74–5
  perception as modular 71
  on prototype theory 108
  proxytype theory 113–22
  sharing concepts 112
propositional attitudes
  beliefs and 2–3
  classical computationalism 35
  human and animal language
    36–7
  language and 2–3
  mental states and 3
  perception 5
  visual processing and 70
prototype theory
  acquiring concepts and 103–4
  concepts and 98
  essentialism 110
  exemplar theory and 104–5
  objections to 105–8
  representations 101–3
proxytype theory
  concepts 98
  objections to 115–19
  Prinz's development of 113–15

Twin Earth thought experiment
  119–2
psychoanalysis and the unconscious
  22–3
psychology
  causal principles 111–12
  cognitive approaches 15
  concept acquisition 105–6
  evolutionary 78
  language and 124–5
  modularity thesis and 72–3
  neuropsychology 72–3
  objects and infants 84–9
  principles of objects 85
  theory theory and 109–11
Pullum, G. K. 142–3
Putnam, Hilary
  Twin Earth thought experiment
    119–2
  type-type identity theory 21–2
Pylyshyn, Z. W. 164

Quine, W. V. O.
  'bootstrapping' 89
  Duhem-Quine thesis 11–13
  hypotheses 76
  'The Two Dogmas of Empiricism'
    11

Radford, A. 133
Ramachandran, V. S. 173–4
reasoning
  classical computationalism and
    37–40
  dual aspect theories 39–40
  scientific 79–81
recursivity 148
reductionism
  Duhem–Quine Theory and 11–13
representation
  anti-representational approaches
    57–62, 64
  computers and 30
  concepts 99
  connectionism and 53
  controversies of 197–8
  the extended mind 62–5
  language and 126, 129–8
  meaning of 40–1
  memory 114
  neuroscience and 163
  philosophy of mind and 17–19
  prototypes and 101–3
  proxytypes and 114, 115
  see also computationalism

Rey, Georges 95
Rizzolatti, G. 166
robots
  cognition and 61
  subsumption architecture 60–1
  Watt machine governors 59–60
Rosch, Eleanor 101
Rumelhart, D. E. 25
Ryle, Gilbert 17

Sampson, G. 142–3
Samuels, R. 79
scans, brain 174–5
Schiffer, Stephen 34–5
Schneider, Susan 31
Scholz, B. C. 142–3
science
  commonsense and 13–14
  empiricism 11–13
  falsifiability of hypotheses 10–11
  natural laws 9–10
  scepticism and 6–9
  theoretical entities 11
  theory theory and 109
  see also the brain; neuroscience
Searle, John 17
senses
  acquiring concepts and 104
  modularity thesis 68–72
  the parietal lobe and 171–2
  phantom limb 173–4
  prior to perception 115
  propositional attitudes and 3
Shapiro, L. 57
Simon, Herbert 79
Simons, D.
  inattentional blindness 181–2,
    193
  multi-tasking and language 140–1
  perception and 70
simulation theory 89–92
Skinner, B. F.
  Chomsky and 24
  stimulus–response learning 15–16
  Verbal Behavior 16, 124
Smart, J. J. C. 162
Smith, E. E. 70
Smith, N. V. 74
Snow, C. E. 144
social structures
  language and 124–5, 138–41
Specific Language Impairment (SLI)
  74
Spelke, E. 84–9
Stenberg, C. 155

Sterelny, K. 80
symbols
 amodal 116
 computer language and 26
 connectionism and 53
 simple and complex 117
Syntactic Structures (Chomsky) 124

teleology, computationalism and
 41
theory theory
 causality 111–12
 concepts and 98
 developmental psychology and
 109–11
 mind-reading 89
 psychological 121
 science and 109
Thinking, Fast and Slow
 (Kahneman) 38–9, 40
thought
 concepts and 94–6
 defining 1–2
 language of thought (LoT) 48
 manifested by language 64–5
 systematicity of 35–6, 55–6
 see also cognition
Tomasello, Michael
 children and language 144, 145
 Chomsky and 150, 154
 Constructing and Language 146
 *Origins of Human
 Communication* 146
 usage-based language acquisition
 146–59
Tooby, J.
 biological mind 78–9
 cheater detection module 81–3
Toribio, J. 62
A Treatise of Human Nature
 (Hume) 14
Tronick, E. 86, 155
Tsimpli, I. M. 74
Turing, Alan
 machines and mind 18–19, 22
 Universal Turing Machine 30
Tversky, Amos 38
Twin Earth thought experiment
 proxytype theory and 119–22

'The Two Dogmas of Empiricism'
 (Quine) 11–13
Tyler, L. K. 74–5
type-type identity theory 20–2
 neuroscience and 162–3

the unconscious
 cognition and 4
 Freud and 22–3
 see also consciousness
Ungerleider, L. G. 177–8
Universal Grammar (UG) 129,
 136–8

Van Gelder, T.
 Watt governor and cognition
 59–60
Verbal Behavior (Skinner) 16, 124
vision
 Bayes's theorem 182–5
 beliefs and perception 186–7
 blindness 178, 179, 181–2, 188
 cognitive science and 175–6
 consciousness and 181–6
 eye and brain function 176–8
 Gibson's theory of 57
 modularity thesis and 68–72
 Noë's enactivism theory 187–94
 perception and 3
 perceptual presence 190
 phenomenology of 192–3
 shapes and 189–90

Wason, P. 81–2, 83
wasp, Sphex 61–2
Watson, J. B. 15
Wernicke, Carl 20
Wiesel, Torsten N. 165, 177
Williams syndrome 73–4
Wilson, R. A. 57
Wittgenstein, Ludwig 196
 games and concepts 100
 images and meaning 100
 mental representation 17
writing, memory and 65
Wundt, Wilhelm 17
Wynn, Karen 85

Zeki, S. 181